Journal of Applied Logics - IfCoLog Journal of Logics and their Applications

Volume 8, Number 10

December 2021

Disclaimer

Statements of fact and opinion in the articles in Journal of Applied Logics - IfCoLog Journal of Logics and their Applications (JALs-FLAP) are those of the respective authors and contributors and not of the JALs-FLAP. Neither College Publications nor the JALs-FLAP make any representation, express or implied, in respect of the accuracy of the material in this journal and cannot accept any legal responsibility or liability for any errors or omissions that may be made. The reader should make his/her own evaluation as to the appropriateness or otherwise of any experimental technique described.

© Individual authors and College Publications 2021
All rights reserved.

ISBN 978-1-84890-381-4
ISSN (E) 2631-9829
ISSN (P) 2631-9810

College Publications
Scientific Director: Dov Gabbay
Managing Director: Jane Spurr

http://www.collegepublications.co.uk

All rights reserved. No part of this publication may be reproduced, stored in a retrieval system or transmitted in any form, or by any means, electronic, mechanical, photocopying, recording or otherwise without prior permission, in writing, from the publisher.

Editorial Board

Editors-in-Chief
Dov M. Gabbay and Jörg Siekmann

Marcello D'Agostino
Natasha Alechina
Sandra Alves
Arnon Avron
Jan Broersen
Martin Caminada
Balder ten Cate
Agata Ciabattoni
Robin Cooper
Luis Farinas del Cerro
Esther David
Didier Dubois
PM Dung
David Fernandez Duque
Jan van Eijck
Marcelo Falappa
Amy Felty
Eduaro Fermé

Melvin Fitting
Michael Gabbay
Murdoch Gabbay
Thomas F. Gordon
Wesley H. Holliday
Sara Kalvala
Shalom Lappin
Beishui Liao
David Makinson
George Metcalfe
Claudia Nalon
Valeria de Paiva
Jeff Paris
David Pearce
Pavlos Peppas
Brigitte Pientka
Elaine Pimentel

Henri Prade
David Pym
Ruy de Queiroz
Ram Ramanujam
Chrtian Retoré
Ulrike Sattler
Jörg Siekmann
Jane Spurr
Kaile Su
Leon van der Torre
Yde Venema
Rineke Verbrugge
Heinrich Wansing
Jef Wijsen
John Woods
Michael Wooldridge
Anna Zamansky

Scope and Submissions

This journal considers submission in all areas of pure and applied logic, including:

pure logical systems	dynamic logic
proof theory	quantum logic
constructive logic	algebraic logic
categorical logic	logic and cognition
modal and temporal logic	probabilistic logic
model theory	logic and networks
recursion theory	neuro-logical systems
type theory	complexity
nominal theory	argumentation theory
nonclassical logics	logic and computation
nonmonotonic logic	logic and language
numerical and uncertainty reasoning	logic engineering
logic and AI	knowledge-based systems
foundations of logic programming	automated reasoning
belief change/revision	knowledge representation
systems of knowledge and belief	logic in hardware and VLSI
logics and semantics of programming	natural language
specification and verification	concurrent computation
agent theory	planning
databases	

This journal will also consider papers on the application of logic in other subject areas: philosophy, cognitive science, physics etc. provided they have some formal content.

Submissions should be sent to Jane Spurr (jane@janespurr.net) as a pdf file, preferably compiled in LaTeX using the IFCoLog class file.

CONTENTS

ARTICLES

Weak Pseudo EMV-algebras. I: Basic Properties 2365
    Anatolij Dvurečenskij and Omid Zahiri

Weak Pseudo EMV-algebras. II: Representation and Subvarieties 2401
    Anatolij Dvurečenskij and Omid Zahiri

A 2 Set-up Routley Semantics for the 4-valued Logic PŁ4 2435
    Gemma Robles and José M. Méndéz

The Embedding Path Order for Lambda-Free Higher-Order Terms 2447
    Alexander Bentkamp

Entailment, Transmission of Truth, and Minimality 2471
    Andrzej Wiśniewski

Weak Pseudo EMV-algebras.
I: Basic Properties

Anatolij Dvurečenskij*
*Mathematical Institute, Slovak Academy of Sciences, Štefánikova 49,
SK-814 73 Bratislava, Slovakia,
Palacký University Olomouc, Faculty of Sciences, tř. 17. listopadu 12,
CZ-771 46 Olomouc, Czech Republic*
dvurecen@mat.savba.sk

Omid Zahiri
Tehran, Iran
zahiri@protonmail.com

Abstract

We define weak pseudo EMV-algebras which are a non-commutative generalization of weak EMV-algebras, pseudo MV-algebras, and generalized Boolean algebras, respectively. In contrast to pseudo EMV-algebras, the class of wPEMV-algebras is a variety. We present basic properties and examples of wPEMV-algebras. The main aim is to show when a wPEMV-algebra can be embedded into a wPEMV-algebra N with top element, called a representing wPEMV-algebra, as a maximal and normal ideal of N. The paper is divided into two parts. Part I studies wPEMV-algebras from the point of semiclans, generalized pseudo effect algebras and integral GMV-algebras. We describe congruences via normal ideals, and we show when a wPEMV-algebra possesses a representing one.

Part II. It studies representable wPEMV-algebras and it shows an equational base for them. Left and right unitizing automorphisms enable us to construct representing wPEMV-algebras. We present the Basic Representation Theorem. Finally, we study subvarieties of cancellative wPEMV-algebras, perfect wPEMV-algebras, weakly commutative wPEMV-algebras, and normal-valued wPEMV-algebras, respectively.

---

*Sponsored by the Slovak Research and Development Agency under contract APVV-16-0073 and the grant VEGA No. 2/0142/20 SAV

1 Introduction

MV-algebras were introduced in 1958, [6, 7], as a many valued counterpart of Boolean algebras. A crucial result on MV-algebras was established in [32] showing that there is a categorical equivalence between the set of MV-algebras forming a variety and the category of unital Abelian ℓ-groups which do not form a variety. Over more than 60 years MV-algebras penetrated in different areas of mathematics and they were generalized in many ways. For more about MV-algebras, see [8]. A non-commutative generalization of MV-algebras was presented in [28] as *pseudo MV-algebras* and independently in [33] as *generalized MV-algebras*. A fundamental generalization of Mundici's representation result for pseudo MV-algebras was presented in [14] showing that the category of pseudo MV-algebras is categorically equivalent to the category of unital ℓ-groups not necessarily Abelian. It is not surprising that during decades we have a whole palette of different generalizations of MV-algebras like BL-algebras, [31], hoops, [3], pseudo hoops, [29], pseudo BL-algebras, [11, 12], residuated lattices [4, 27], effect algebras, [25], pseudo effect algebras, [15, 16], semiclans, [5], BCK-algebras, Wajsberg hoops, [1], etc.

In [19, 21, 22], we have introduced a commutative and non-commutative generalization of both MV-algebras and generalized Boolean algebras as algebras where top element is not assumed a priori and every element is dominated by some idempotent element, moreover, every interval $[0, a]$, where a is an idempotent, is an MV-algebra or a pseudo MV-algebra. These algebras are said to be *EMV-algebras* or *pseudo EMV-algebras* (EMV stands for extended MV-algebras). If such an algebra possesses a top element, then it is equivalent to an MV-algebra or to a pseudo MV-algebra. The principal representing result says that every EMV-algebra (pseudo EMV-algebra) M without top element can be embedded into an EMV-algebra (pseudo EMV-algebra) N with top element as a maximal and normal ideal of N and every element not belonging to the image of M is a complement of the image of some element from M. A variant of the Loomis–Sikorski theorem for EMV-algebras was established in [20].

We have to underline that not every maximal ideal of any MV-algebra can serve as an example of EMV-algebras. This is the case, e.g., for the Chang MV-algebra $\Gamma(\mathbb{Z} \overrightarrow{\times} \mathbb{Z}, (1,0))$, where the unique maximal ideal is the set $M = \{(0, n) \mid n \geq 0\}$ and it has only one idempotent $(0, 0)$ which does not dominate all elements of M. We note that $\overrightarrow{\times}$ denotes the lexicographic product. If we compare EMV-algebras with Wajsberg hoops, that are bottom-free subreducts of MV-algebras, we can find examples of Wajsberg hoops, [1, Prop 2.1], [3, Ex 1.5], which are not EMV-algebras and vice-versa.

Neither EMV-algebras nor pseudo EMV-algebras form a variety because they are not closed under subalgebras, but they are close to varieties, special classes called q-varieties. Consequently, there are countably many q-subvarieties of EMV-algebras and uncountably many q-subvarieties of pseudo EMV-algebras.

Therefore, in [23], we have found classes of algebras called *weak EMV-algebras* (wEMV-algebras) which form a variety and this class contains also all EMV-algebras and this variety is the least variety containing all wEMV-algebras.

In this paper we present a non-commutative generalization of wEMV-algebras called *weak pseudo EMV-algebras* (wPEMV-algebras in short). They are algebras $(M; \vee, \wedge, \oplus, \ominus, \oslash, 0)$ of type $(2, 2, 2, 2, 2, 0)$ and they form a variety.

The main goals of the paper, which is divided into two parts, are:

Part I.

(1) We investigate the basic properties of wPEMV-algebras and present basic examples of wPEMV-algebras.

(2) Show that every wPEMV-algebra can be embedded into the positive cone of some ℓ-group preserving $\vee, \wedge, \ominus, \oslash$ and partial addition. In addition, we show a relation of wPEMV-algebras with generalized pseudo effect algebras. Moreover, every interval $[0, c]$ in M can be converted into a pseudo MV-algebra such that it is not necessarily a subalgebra of the original wPEMV-algebra M.

(3) Study ideals, normal ideals, congruences, and show that a wPEMV-algebra M without top element can be embedded into a wPEMV-algebra N with top element as a maximal and normal ideal of N and every element of N not belonging to the image of M is a complement of some element from the image of M. This wPEMV-algebra N is said to be *representing* M. First such a study for generalized Boolean algebras was done in [9] and for EMV-algebras, pseudo EMV-algebras and wEMV-algebras in [19, 22, 23].

Part II.

(4) Study representable wPEMV-algebras, i.e. wPEMV-algebras that are subdirect product of linearly ordered wPEMV-algebras. We show that every representable wPEMV-algebra admits a representable and representing wPEMV-algebra. These algebras form a variety and we present an equational base for them. We show that this variety contains uncountably many subvarieties.

(5) We introduce unitizing left and right automorphisms of wPEMV-algebras and we show that the existence of a left (right) unitizing automorphism is a necessary and sufficient condition in order that a wPEMV-algebra does possess a representing wPEMV-algebra. First we show that associated wPEMV-algebras, cancellative wPEMV-algebras, commutative or weakly commutative wPEMV-algebras have representing wPEMV-algebras. We present the Basic Representation Theorem for wPEMV-algebras which entails that unitizing automorphisms always exist in each wPEMV-algebra. Therefore, every wPEMV-algebra embeds in a representing one.

(6) We study special kinds of subvarieties of wPEMV-algebras. For example, the subvariety of cancellative wPEMV-algebras with an equational base, perfect wPEMV-algebras (and we show its equational base), normal-valued wPEMV-algebras and we present some atoms in the lattice of subvarieties.

Some open questions are formulated.

The paper is organized as follows. Part I. Section 2 presents the basic properties of pseudo MV-algebras and EMV-algebras. In Section 3, we introduce weak pseudo EMV-algebras with basic properties and show important examples of wPEMV-algebras. In Section 4, we define a partial addition + and show how it is connected with semiclans in the sense of [5] and with generalized pseudo effect algebras. Moreover, we show that wPEMV-algebras can be viewed also as integral GMV-algebras in the sense of [2, 27, 4], and vice-versa. Congruences, ideals, normal ideals are studied in Section 5 and we show that a wPEMV-algebra without top element can be embedded into a wPEMV-algebra with top element.

Part II. In Section 6, we study in detail representable wPEMV-algebras, we show that they can be always embedded into a representable wPEMV-algebra with top element. We present an equational base for this variety. Section 7 is devoted to a question when a wPEMV-algebra without top element is a maximal and normal ideal of some wPEMV-algebra with top element. Therefore, we study left and right unitizing automorphisms. The main result, the Basic Representing Theorem, will be presented. Final Section 8 describes different subvarieties of wPEMV-algebras: cancellative, perfect, weakly commutative and normal-valued ones.

2 Pseudo MV-algebras and Pseudo EMV-algebras

In the section, we present basic properties of pseudo MV-algebras and pseudo EMV-algebras which are non-commutative generalizations of MV-algebras and EMV-algebras, respectively.

We note that pseudo EMV-algebras generalize pseudo MV-algebras which were defined in [28] and equivalently in [33] as a generalized MV-algebras which are a non-commutative extension of MV-algebras. Thus due to [28], a *pseudo MV-algebra* is an algebra $(M; \oplus, ^-, ^\sim, 0, 1)$ of type $(2, 1, 1, 0, 0)$ such that the following axioms hold for all $x, y, z \in M$ with an additional binary operation $\odot$ defined via

$$y \odot x = (x^- \oplus y^-)^\sim$$

(A1) $x \oplus (y \oplus z) = (x \oplus y) \oplus z$;

(A2) $x \oplus 0 = 0 \oplus x = x$;

(A3) $x \oplus 1 = 1 \oplus x = 1$;

(A4) $1^\sim = 0$; $1^- = 0$;

(A5) $(x^- \oplus y^-)^\sim = (x^\sim \oplus y^\sim)^-$;

(A6) $x \oplus (x^\sim \odot y) = y \oplus (y^\sim \odot x) = (x \odot y^-) \oplus y = (y \odot x^-) \oplus x$;

(A7) $x \odot (x^- \oplus y) = (x \oplus y^\sim) \odot y$;

(A8) $(x^-)^\sim = x$.

Pseudo MV-algebras are closely connected with unital ℓ-groups. We note that an element u from the positive cone G^+ of an ℓ-group G is said to be a *strong unit* if given $x \in G$, there is an integer $n \geq 0$ such that $x \leq nu$. A couple (G, u), where u is a fixed strong unit of G, is said to be a *unital ℓ-group*. For more info on ℓ-groups, we recommend to consult [10, 26, 30].

For example, if (G, u) is a unital ℓ-group (not necessarily Abelian), we define

$$\Gamma(G, u) := [0, u]$$

and

$$x \oplus y := (x + y) \wedge u,$$
$$x^- := u - x,$$
$$x^\sim := -x + u,$$
$$x \odot y := (x - u + y) \vee 0,$$

then $\Gamma(G, u) := ([0, u]; \oplus, ^-, ^\sim, 0, u)$ is a pseudo MV-algebra [28]. A basic result on representation of pseudo MV-algebras says that there is a one-to-one correspondence between pseudo MV-algebras and unital ℓ-groups, for more details see [14].

Pseudo EMV-algebras were introduced in [21]:

Definition 2.1. An algebra $(M; \vee, \wedge, \oplus, 0)$ of type $(2, 2, 2, 0)$ is called a *pseudo EMV-algebra* if it satisfies the following conditions:

(E1) $(M; \vee, \wedge, 0)$ is a distributive lattice with the least element 0;

(E2) $(M; \oplus, 0)$ is an ordered monoid with a neutral element 0;

(E3) for each $a \in \mathcal{I}(M) := \{x \in M : x \oplus x = x\}$, the elements

$$\lambda_a(x) := \min\{z \in [0, a] : z \oplus x = a\}, \quad \rho_a(x) := \min\{z \in [0, a] : x \oplus z = a\}$$

exist in M for all $x \in [0, a]$, and the algebra $([0, a]; \oplus, \lambda_a, \rho_a, 0, a)$ is a pseudo MV-algebra;

(E4) for each $x \in M$, there is $a \in \mathcal{I}(M)$ such that $x \leq a$.

It is noteworthy of recalling that the orders following from (E1) and (E2), respectively, are the same.

In an analogy with pseudo MV-algebras, we can define a total binary operation $\odot$ in the following way: For all $x, y \in M$, we define

$$x \odot y = \rho_a(\lambda_a(y) \oplus \lambda_a(x)),$$

where $a \in \mathcal{I}(M)$ and $x, y \in [0, a]$. Then $x \odot y$ is correctly defined and it does not depend on $a \in \mathcal{I}(M)$, and

$$x \odot y = \lambda_a(\rho_a(y) \oplus \rho_a(x)).$$

In addition, if $x, y \in M$, $x \leq y$, then

$$y \odot \lambda_a(x) = y \odot \lambda_b(x), \quad \rho_a(x) \odot y = \rho_b(x) \odot y \tag{2.1}$$

for all idempotents a, b of M with $x, y \leq a, b$, and

$$y = (y \odot \lambda_a(x)) \oplus x = x \oplus (\rho_a(x) \odot y). \tag{2.2}$$

If $x, y \in [0, a]$ for some idempotent $a \in M$, then

$$x \odot \lambda_a(y) = x \odot \lambda_a(x \wedge y), \quad \rho_a(y) \odot x = \rho_a(x \wedge y) \odot x, \tag{2.3}$$

and

$$(x \odot \lambda_a(y)) \oplus (x \wedge y) = x = (x \wedge y) \oplus (\rho_a(y) \odot x). \tag{2.4}$$

Finally, if $x, y \leq a \in \mathcal{I}(M)$, then

$$((x \oplus y) \odot \lambda_a(x)) \oplus x = x \oplus y = y \oplus (\rho_a(y) \odot (x \oplus y)) \tag{2.5}$$

and if $x \leq a \in \mathcal{I}(M)$, then

$$x \odot \lambda_a(x) = 0 = \rho_a(x) \odot x. \tag{2.6}$$

It is easy to verify that if $(M; \oplus, ^-, ^\sim, 0, 1)$ is a pseudo MV-algebra, then $(M; \vee, \wedge, \oplus, 0)$ is a pseudo EMV-algebra with top element. Conversely, if $(M; \vee, \wedge, \oplus, 0)$ is a pseudo EMV-algebra with top element 1, then $(M; \oplus, \lambda_1, \rho_1, 0, 1)$ is a pseudo MV-algebra. This shows that there is a one-to-one correspondence between pseudo MV-algebras and pseudo EMV-algebras with top element.

The basic representation theorem for pseudo EMV-algebras was established in [22, Thm 6.4]:

Theorem 2.2. [Basic Representation Theorem] *Every pseudo EMV-algebra M is either a pseudo EMV-algebra with top element or M can be embedded into a pseudo EMV-algebra N with top element as a maximal and normal ideal of N such that every element $x \in N$ is either from the image of M or there is a unique x_0 from the image of M such that $x = \rho_1(x_0)$.*

We remind that a subset I of a pseudo EMV-algebra is an *ideal* if I is closed under $\oplus$ and if $x \leq y \in I$, then $x \in I$. An ideal I is *normal* if $x \oplus I = I \oplus x$ for each $x \in M$.

Remark 2.3. Let $(M; \vee, \wedge, \oplus, 0)$ be a pseudo EMV-algebra and $x, y, z \in M$ such that $x \odot y \leq z$. Choose $a \in \mathcal{I}(M)$ with $x, y \leq a$. Then by [21, Prop 3.4], $x \odot y = x \odot_a y$, where $\odot_a$ is the binary operation on the pseudo MV-algebra $([0,a]; \oplus, \lambda_a, \rho_a, 0)$, that is $s \odot_a t = \lambda_a(\rho_a(t) \oplus \rho_a(s))$ for all $s, t \in [0, a]$. By [28] in $[0, a]$. We have

(i) $x \odot_a y \leq z$ if and only if $x \odot_a y \leq z \wedge a$ if and only if $x \odot_a \lambda_a(\rho_a(y)) \leq z \wedge a$ if and only if $x \leq (z \wedge a) \oplus \rho_a(y) = z \oplus \rho_a(y)$.

(ii) $x \odot_a y \leq z$ if and only if $x \odot_a y \leq z \wedge a$ if and only if $\rho_a(\lambda_a(x)) \odot_a y \leq z \wedge a$ if and only if $y \leq \lambda_a(x) \oplus (z \wedge a) = \lambda_a(x) \oplus z$.

Lemma 2.4. *Let $(M; \vee, \wedge, \oplus, 0)$ be a pseudo EMV-algebra. For each $x, y \in M$, we define*

$$x \ominus y = x \odot \lambda_a(y), \quad \text{where } x, y \leq a \in \mathcal{I}(M), \qquad (2.7)$$
$$x \oslash y = \rho_a(x) \odot y, \quad \text{where } x, y \leq a \in \mathcal{I}(M). \qquad (2.8)$$

Then $\ominus : M \times M \to M$ and $\oslash : M \times M \to M$ are well-defined binary operations on M which do not depend on $a \in \mathcal{I}(M)$ if $x, y \leq a$.

Proof. Let $x, y \in M$ and $a, b \in \mathcal{I}(M)$ be such that $x, y \leq a, b$. Choose $c \in \mathcal{I}(M)$ with $a, b \leq c$. Since $[0, c]$ is a pseudo EMV-algebra, by [21, Prop 3.3(ii)] we have

$$x \odot \lambda_a(y) = x \odot (\lambda_c(y) \wedge a) = (x \odot \lambda_c(y)) \wedge (x \odot a), \text{ by [28, Prop 1.22]}$$
$$= (x \odot \lambda_c(y)) \wedge x = x \odot \lambda_c(y), \text{ since } x \odot \lambda_c(y) \leq x.$$

In a similar way, we can show that $x \odot \lambda_b(y) = x \odot \lambda_c(y)$. That is, $x \odot \lambda_a(y) = x \odot \lambda_b(y)$. Therefore, $\ominus$ is well-defined.

A similar proof works for $\oslash$ instead of $\ominus$. $\square$

3 Weak Pseudo EMV-algebras

In [23], we have introduced weak EMV-algebras $(M; \vee, \wedge, \oplus, \ominus, 0)$ that are algebras with a commutative operation $\oplus$. A non-commutative generalization of wEMV-algebras are weak pseudo EMV-algebras:

Definition 3.1. An algebra $(M; \vee, \wedge, \oplus, \ominus, \oslash, 0)$ of type $(2,2,2,2,2,0)$ is called a *wPEMV-algebra* (w means weak) if it satisfies the following conditions:

(W1) $(M, \vee, \wedge, 0)$ is a distributive lattice with the least element 0;

(W2) $(M; \oplus, 0)$ is a monoid;

(W3) $(y \oplus x) \ominus x \leq y$ and $x \oslash (x \oplus y) \leq y$;

(W4) $(y \ominus x) \oplus x = x \vee y = x \oplus (x \oslash y)$;

(W5) $x \ominus (x \wedge y) = x \ominus y$ and $(x \wedge y) \oslash y = x \oslash y$;

(W6) $y \ominus (x \oslash y) = x \wedge y = (y \ominus x) \oslash y$;

(W7) $z \ominus (x \vee y) = (z \ominus x) \wedge (z \ominus y)$ and $(x \vee y) \oslash z = (x \oslash y) \wedge (y \oslash z)$;

(W8) $(x \wedge y) \ominus z = (x \ominus z) \wedge (y \ominus z)$ and $z \oslash (x \wedge y) = (z \oslash x) \wedge (z \oslash y)$;

(W9) $x \ominus (y \oplus z) = (x \ominus z) \ominus y$ and $(y \oplus z) \oslash x = z \oslash (y \oslash x)$;

(W10) $x \oplus (y \vee z) = (x \oplus y) \vee (x \oplus z)$ and $(y \vee z) \oplus x = (y \oplus x) \vee (z \oplus x)$.

If $(M; \vee, \wedge, \oplus, \ominus, 0)$ is a wEMV-algebra, then $(M; \vee, \wedge, \oplus, \ominus, \oslash, 0)$, where $x \oslash y = y \ominus x$, $x, y \in M$, is a wPEMV-algebra. Conversely, if $(M; \vee, \wedge, \oplus, \ominus, \oslash, 0)$ is a wPEMV-algebra with commutative $\oplus$, then $(M; \vee, \wedge, \oplus, \ominus, 0)$ is a wEMV-algebra; see also Proposition 3.2(xv) below.

An element a of a wPEMV-algebra M is said to be an *idempotent* (or a *Boolean element*) if $a \oplus a = a$. We recall that $\mathcal{I}(M)$ is the set of idempotents of M. Then $0 \in \mathcal{I}(M)$.

We say that a monoid $(M; \oplus, 0)$ endowed with a partial order $\leq$ is (i) *ordered* if $x \leq y$ implies $z_1 \oplus x \oplus z_2 \leq z_1 \oplus y \oplus z_2$ for all $z_1, z_2 \in M$, (ii) *right naturally ordered* if $x \leq y$ iff there is $v \in M$ such that $y = x \oplus v$, (iii) *left naturally ordered* if $x \leq y$ iff there is $u \in M$ such that $y = u \oplus x$, and (iv) *naturally ordered* if $x \leq y$ iff there are $u, v \in M$ such that $x \oplus v = y = u \oplus x$. Given $x \in M$ and any integer $n \geq 0$, we define

$$0.x := 0, \quad (n+1).x := n.x \oplus x, \quad \text{if } n \geq 0.$$

The basic properties of wPEMV-algebras are as follows.

Proposition 3.2. *Let $(M; \vee, \wedge, \oplus, \ominus, \oslash, 0)$ be a wPEMV-algebra and $x, y, z \in M$. Then the following hold:*

(i) $(M; \oplus, 0)$ is an ordered monoid which is right and left naturally ordered (that is, $x \leq y$ if and only if there is $u \in M$ such that $x \oplus u = y$, equivalently, there is $v \in M$ such that $v \oplus x = y$).

(ii) $(a \ominus x) \oslash a = x = a \ominus (x \oslash a)$ if $x \leq a$.

(iii) $x \wedge y = ((a \ominus x) \vee (a \ominus y)) \oslash a$ and $x \wedge y = a \ominus ((x \oslash a) \vee (y \oslash a))$ if $x, y \leq a$.

(iv) $x \leq y$ implies that $x \ominus z \leq y \ominus z$ and $z \oslash x \leq z \oslash y$. Also, $z \ominus y \leq z \ominus x$ and $y \oslash z \leq x \oslash z$.

(v) $z \leq x \oplus y$ if and only if $z \ominus y \leq x$ if and only if $x \oslash z \leq y$.

(vi) $(x \wedge y) \oplus z = (x \oplus z) \wedge (y \oplus z)$ and $z \oplus (x \wedge y) = (z \oplus x) \wedge (z \oplus y)$.

(vii) $z \ominus (x \wedge y) = (z \ominus x) \vee (z \ominus y)$ and $(x \wedge y) \oslash z = (x \oslash z) \vee (y \oslash z)$.

(viii) $x \ominus x = 0 = x \oslash x$ and $x \ominus 0 = x = 0 \oslash x$.

(ix) $x \leq y$ if and only if $x \ominus y = 0$ if and only if $y \oslash x = 0$.

(x) $(x \vee y) \ominus z = (x \ominus z) \vee (y \ominus z)$ and $z \oslash (x \vee y) = (z \oslash x) \vee (z \oslash y)$.

(xi) $x \ominus y \leq x$ and $x \oslash y \leq y$.

(xii) $(x \ominus y) \wedge (y \ominus x) = 0 = (x \oslash y) \wedge (y \oslash x)$.

(xiii) If $a \oplus a = a$, then $a \oplus x = a \vee x = x \oplus a$.

(xiv) $z \ominus (x \wedge y) = (z \ominus x) \vee (z \ominus y)$ and $(x \wedge y) \oslash z = (x \oslash z) \vee (y \oslash z)$.

(xv) The binary operation $\oplus$ is commutative if and only if $x \ominus y = y \oslash x$.

Proof. (i) By (W10), $(M; \oplus, 0)$ is an ordered monoid. We show that M is left naturally ordered. Let $x, y \in M$ such that $x \leq y$. Then $y = x \vee y = (y \ominus x) \oplus x$. Set $v = y \ominus x$. Conversely, let $y = v \oplus x$ for some $v \in M$. Since M is ordered, then $x = 0 \oplus x \leq v \oplus x = y$. Whence, M is left naturally ordered.

In a similar way, $x \leq y$ iff there is $u \in M$ such that $y = x \oplus u$.

(ii) By (W6), $(a \ominus x) \obar a = x \wedge a = x$ and $a \ominus (x \obar a) = a \wedge x = x$.

(iii) By (ii), $x \wedge y = (a \ominus (x \obar a)) \wedge (a \ominus (y \obar a))$. So, by (W7),
$$x \wedge y = (a \ominus (x \obar a)) \wedge (a \ominus (y \obar a)) = a \ominus ((x \obar a) \vee (y \obar a)).$$

The proof of the second part is similar.

(iv) It follows from (W7)–(W8).

(v) If $z \leq x \oplus y$, then by (iv) and (W3), $z \ominus y \leq (x \oplus y) \ominus y \leq x$. Conversely, if $z \ominus y \leq x$, then by (W4), $z \leq z \vee y = (z \ominus y) \oplus y \leq x \oplus y$. In a similar way by (iv) and (W3), we get that $x \obar z \leq x \obar (x \oplus y) \leq y$. Conversely, if $(x \obar z) \leq y$, then by (W4), $z \leq x \vee z = x \oplus (x \obar z) \leq x \oplus y$.

(vi) By (i), $(x \wedge y) \oplus z \leq x \oplus z, y \oplus z$. Let $t \leq x \oplus z, y \oplus z$. From (v) it follows that $t \ominus z \leq x, y$, so that $t \ominus z \leq x \wedge y$ and so by (W4), $t \leq (t \ominus z) \oplus z \leq (x \wedge y) \oplus z$. Hence, $(x \wedge y) \oplus z = (x \oplus z) \wedge (y \oplus z)$. The proof of the second part is similar.

(vii) By (iv), $z \ominus (x \wedge y) \geq z \ominus x, z \ominus y$. Let $z \ominus x, z \ominus y \leq u \in M$. Then by (v), $z \leq u \oplus x, u \oplus y$ and so by (W10), we obtain $z \leq (u \oplus x) \wedge (u \oplus y) = u \oplus (x \wedge y)$ (by (vi)). It follows from (v) that $z \ominus (x \wedge y) \leq u$. Therefore, $z \ominus (x \wedge y) = (z \ominus x) \vee (z \ominus y)$. The proof of the second part is analogous.

(viii) We have $0 \leq x \ominus x = (0 \oplus x) \ominus x \leq 0$ when we have used (W3). Dually we prove $x \obar x = 0$.

Now, we have $x \leq x = x \oplus 0$, so that by (v), $x \ominus 0 \leq x$. On the other hand, $x \ominus 0 \leq x \ominus 0$ and again by (v), $x \leq (x \ominus 0) \oplus 0 = x \ominus 0$, which yields $x \ominus 0 = x$. In a dual way, we prove $0 \obar x = x$.

(ix) If $x \leq y$, then $y = x \vee y$ and so by (iv) and (viii), $x \ominus y \leq (x \vee y) \ominus y = y \ominus y = 0$ and $y \obar x \leq y \obar (x \vee y) = y \obar y = 0$.

Conversely, let $x \ominus y = 0$. By (W4), $y = (x \ominus y) \oplus y = x \vee y$, so that $x \leq y$. In a similar way we prove the second equivalence.

(x) By (iv), we have $(x \vee y) \ominus z \geq (x \ominus z) \vee (y \ominus z)$. Let $c \geq x \ominus z, y \ominus z$. (v) implies $c \oplus z \geq x, y$ and $c \oplus z \geq x \vee y$ which yields $c \geq (x \vee y) \ominus z$. We establish the second property in the dual way.

(xi) Let $x, y \in M$. By (W10), $x \leq x \oplus y$ and $y \leq x \oplus y$. It follows from (v) that $x \ominus y \leq x$ and $x \obar y \leq y$.

(xii) By (W5) and (W8), $(x \ominus y) \wedge (y \ominus x) = (x \ominus (x \wedge y)) \wedge (y \ominus (x \wedge y)) = (x \wedge y) \ominus (x \wedge y) = 0$. A similar proof works for the second part.

(xiii) Using (W4) and (W7), we have $a \vee x = (x \ominus a) \oplus a = (x \ominus a) \oplus a \oplus a = (x \vee a) \oplus a = (x \oplus a) \vee (a \oplus a) = (x \oplus a) \vee a = x \oplus a$. In a dual way, we establish $a \vee x = a \oplus (a \oslash x) = a \oplus (a \vee x) = a \vee (a \oplus x) = a \oplus x$.

(xiv) Due to (iv), we have $z \ominus (x \wedge y) \geq z \ominus x, z \ominus y$. Let $a \geq z \ominus x, z \ominus y$. Then (v) entails $a \oplus x \geq z$ and $a \oplus y \geq z$, so that $x, y \geq a \oslash z$ and $x \vee y \geq a \oslash z$, $a \oplus (x \vee y) \geq z$ and finally, $a \geq z \ominus (x \vee y)$ which finishes the proof. In the dual way we prove the second equality.

(xv) Let $\oplus$ be commutative. Using (W4), we have

$$x = (x \wedge y) \oplus ((x \wedge y) \oslash x) = (x \wedge y) \oplus (y \oslash x) = (y \oslash x) \oplus (x \wedge y),$$
$$x = (x \ominus (x \wedge y)) \oplus (x \wedge y) = (x \ominus y) \oplus (x \wedge y) = (x \wedge y) \oplus (x \ominus y).$$

Therefore, using Proposition 3.2(v), we have

$$y \oslash x = (x \wedge y) \oslash x \leq x \ominus y = x \ominus (x \wedge y) \leq y \oslash x.$$

Conversely, let $x \ominus y = y \oslash x$ for all $x, y \in M$. By (W4), we have

$$((x \oplus y) \ominus x) \oplus x = x \oplus y$$
$$(x \oslash (x \oplus y)) \oplus x = x \oplus y$$
$$y \oplus x \geq x \oplus y \quad \text{(by (W3))}.$$

In the same way, we have $x \oplus y \geq y \oplus x$, yielding $x \oplus y = y \oplus x$. $\square$

Now, we present three kinds of wPEMV-algebras.

Example 3.3. Let $(M; \vee, \wedge, \oplus, 0)$ be a pseudo EMV-algebra and $\ominus$ and $\oslash$ be the binary operations defined in Lemma 2.4. Then $(M; \vee, \wedge, \oplus, \ominus, \oslash, 0)$ is a wPEMV-algebra.

Proof. Clearly, (W1) and (W2) hold. First, we note that, for each $a \in \mathcal{I}(M)$, $([0, a]; \oplus, \lambda_a, \rho_a, 0)$ is a pseudo MV-algebra. Let $x, y \in M$. There is $a \in \mathcal{I}(M)$ such that $x, y \leq a$. Then by Lemma 2.4 and [28, Prop 1.13], we have

$$x \vee y = x \oplus (x \oslash y) = y \oplus (y \oslash x) = (y \ominus x) \oplus x = (x \ominus y) \oplus y, \quad (3.1)$$
$$x \wedge y = x \odot (\lambda_a(x) \oplus y) = y \odot (\lambda_a(y) \oplus x) \quad (3.2)$$
$$= (y \oplus \rho_a(x)) \odot x = (x \oplus \rho_a(y)) \odot y. \quad (3.3)$$

(W3) $(y \oplus x) \ominus x = (y \oplus x) \odot \lambda_a(x) = y \wedge \lambda_a(x) \leq y$. Also, $x \oslash (x \oplus y) \leq y = \rho_a(x) \odot (x \oplus y) = \rho_a(x) \wedge y \leq y$.
(W4) Follows from (3.1).

(W5) Since $([0,a]; \oplus, \lambda_a, \rho_a, 0)$ is a pseudo MV-algebra, by [28, Prop 1.16], $x \ominus (x \wedge y) = x \odot \lambda_a(x \wedge y) = x \odot (\lambda_a(x) \vee \lambda_a(y)) = (x \odot \lambda_a(x)) \vee (x \odot \lambda_a(y)) = x \ominus y$. Moreover, $(x \wedge y) \obslash y = \rho_a(x \wedge y) \odot y = (\rho_a(x) \odot y) \vee (\rho_a(y) \odot y) = \rho_a(x) \odot y = x \obslash y$.
(W6) $y \ominus (x \obslash y) = y \odot \lambda_a(\rho_a(x) \odot y) = y \odot (\lambda_a(y) \oplus x) = x \wedge y$ and $(y \ominus x) \obslash y = \rho_a(y \odot \lambda_a(x)) \odot y = (x \oplus \rho_a(y)) \odot y = x \wedge y$.
(W7) Let $b \in \mathcal{I}(M)$ such that $x, y, z \leq b$. Since $([0,b]; \oplus, \lambda_b, \rho_b, 0)$ is a pseudo MV-algebra, by definition of $\odot$ and using [28, Prop 1.22], we have $z \ominus (x \vee y) = z \odot \lambda_b(x \vee y) = z \odot (\lambda_b(x) \wedge \lambda_b(y)) = (z \odot \lambda_b(x)) \wedge (z \odot \lambda_b(y)) = (z \ominus x) \wedge (z \ominus y)$. In a similar way, we can prove the second part.
(W8) The proof is similar to the proof of (W7).
(W9) $(x \ominus z) \ominus y = (x \odot \lambda_b(z)) \odot \lambda_b(y) = x \odot (\lambda_b(z) \odot \lambda_b(y)) = x \odot \lambda_b(y \oplus z) = x \ominus (y \oplus z)$ and $z \obslash (y \obslash x) = \rho_b(z) \odot (\rho_b(y) \odot x) = (\rho_b(z) \odot \rho_b(y)) \odot x = \rho_b(y \oplus z) \odot x = (y \oplus z) \obslash x$.
(W10) It follows from [28, Prop 1.21]. □

Assume that $(M; \vee, \wedge, \oplus, \ominus, \obslash, 0)$ is a wPEMV-algebra such that its reduct $(M; \vee, \wedge, \oplus, 0)$ is a pseudo EMV-algebra. Then the wPEMV-algebra M is said to be an *associated wPEMV-algebra*. In Example 3.3, we have seen that, for each pseudo EMV-algebra $(M; \vee, \wedge, \oplus, 0)$, we can built the associated wPEMV-algebra $(M; \vee, \wedge, \oplus, \ominus, \obslash, 0)$. We can easily prove that there is a one-to-one correspondence between associated wPEMV-algebras and pseudo EMV-algebras. Let wPEMV, PEMV, and wPEMV$_a$ be the classes of wPEMV-algebras, pseudo EMV-algebras, and associated wPEMV-algebras, respectively.

Example 3.4. Let $(M; \oplus, ^-, ^\sim, 0, 1)$ be a pseudo MV-algebra. Then $(M; \oplus, \vee, \wedge, \ominus, \obslash, 0)$, where $x \ominus y = x \odot y^-$ and $x \obslash y = x^\sim \odot y$, is a wPEMV-algebra with top element 1.

If $\Gamma(G, u)$ is a pseudo MV-algebra determined by a unital ℓ-group (G, u), we denote by $\Gamma_a(G, u)$ the associated wPEMV-algebra corresponding to the pseudo MV-algebra $\Gamma(G, u)$ according to Example 3.4.

Example 3.5. Let G^+ be the positive cone of an ℓ-group G. Set $x \oplus y = x+y$, $x \ominus y = (x-y) \vee 0$, and $x \obslash y = (-x+y) \vee 0$ for all $x, y \in G^+$. Then $(G^+; \vee, \wedge, \oplus, \ominus, \obslash, 0)$ is a wPEMV-algebra called a *wPEMV-algebra of a positive cone* or a *conical wPEMV-algebra*.

4 wPEMV-algebras, Semiclans and Generalized Pseudo Effect algebras

Comparing wPEMV-algebras with semiclans, we show that for each element a of a wPEMV-algebra M, the interval $[0, a] = \{x \in M \mid 0 \leq x \leq a\}$ can be converted into a wPEMV-algebra $M_a = ([0,a]; \vee_a, \wedge_a, \oplus_a, \ominus_a, \oslash_a, 0)$ with top element, where $x \oplus_a y = (x \oplus y) \wedge a$, $x, y \in M$, and $\vee_a, \wedge_a, \ominus_a, \oslash_a$ are restrictions of $\vee, \wedge, \ominus, \oslash$ to $[0,a] \times [0,a]$. We show that every M can be embedded into the positive cone of some ℓ-group preserving $\vee, \wedge, \ominus, \oslash, +$. In addition, we present how a wPEMV-algebra M can be converted into a GPEA-algebra with a special kind of the Riesz Decomposition Property. Moreover, taking into account integral GMV-algebras, a special class of residuated lattices, we show that wPEMV-algebras are equivalent to integral GMV-algebras.

According to [5], we introduce the following important notion. We say that a partial algebra $(C; \wedge, +)$ is a *semiclan* if it is a $\wedge$-semilattice and a partial algebra with respect to $+$ such that the following axioms are satisfied:

(S1) if $a \leq b$, then there exist $x, y \in C$ such that $b = a + x$ and $b = y + a$;

(S2) if $a + x$, $a + y \in C$, $a + x = a + y$, then $x = y$, and if $x + a$, $y + a \in C$, $x + a = y + a$, then $x = y$;

(S3) if $a + x$, $a + y \in C$, then $(a + x) \wedge (a + y) = a + (x \wedge y)$, and if $x + a$, $y + a \in C$, then $(x + a) \wedge (y + a) = (x \wedge y) + a$;

(S4) $a + b$, $(a + b) + c \in C$ iff $b + c$, $a + (b + c) \in C$, and in this case we have $(a + b) + c = a + (b + c)$;

(S5) if $(a \wedge b) + c = c$ and $a \vee b$ exists, then $a + b = a \vee b = b + a$.

It is clear that if G is an ℓ-group, then $(G^+; \wedge, +)$ is a semiclan, and by [5, p. 321], for any semiclan $(C; \wedge, +)$ there exists an ℓ-group G with the positive cone G^+ such that C can be embedded into the semiclan $(G^+; \wedge, +)$ preserving $+$, $\wedge$, and all existing $\vee$ in C. In addition, every semiclan contains a neutral element 0, that is, $x + 0, 0 + x$ exist and $x + 0 = x = 0 + x$, [5, (1.1), p. 317].

Now, fix an arbitrary element c from a wPEMV-algebra M and define the interval $[0, c] = \{x \in M \mid 0 \leq x \leq c\}$. On the interval $[0, c]$, we define a partial operation $+ = +_c$ which is defined as follows: For $x, y \in [0, c]$, $x + y$ is defined iff $x \leq \lambda_c(y)$, and in such a case, $x + y := x \oplus y$. The basic properties of the partial addition $+$ on $[0, c]$ are as follows.

Lemma 4.1. *Let $c \in M$ be fixed and $x, y \in [0, c]$. Then:*

(i) $x + y$ *is defined if and only if* $y \leq \rho_c(x)$.

(ii) *If* $x + y$ *is defined, then* $x + y \in [0, c]$.

(iii) $x + 0$ and $0 + x$ always exist and $x + 0 = x = 0 + x$.

(iv) If $x + y$ is defined in $[0, c]$ and $x_1 \leq x$ and $y_1 \leq y$, then $x_1 + y_1$ is defined in $[0, c]$.

(v) $(c \ominus x) + x$ and $x + (x \oslash c)$ exist in $[0, c]$, and $(c \ominus x) + x = c = x + (x \oslash c)$.

(vi) If $\lambda_c(x) + y = c$, then $x = y$. If $y + \rho_c(x) = c$, then $x = y$.

(vii) $(x + y) \ominus y = x$ and $x \oslash (x + y) = y$.

Proof. (i) It follows from the fact $x \leq \lambda_c(y)$ implies $y \leq \rho_c(x)$ (use Proposition 3.2(ii),(iv)).

(ii) The inequality $x \leq \lambda_c(y)$ yields $x \oplus y \leq \lambda_c(y) \oplus y = c$.

(iii) It is evident.

(iv) In any rate, $x_1 \leq x \leq \lambda_c(y) \leq \lambda_c(y_1)$.

(v) This follows from (W4).

(vi) Using (W3), we have $\lambda_c(y) = c \ominus y = (\lambda_c(x) + y) \ominus y \leq \lambda_c(x) \leq \lambda_c(y)$, so that $x = y$. In a similar way, we have the second property.

(vii) Since $(x + y) \ominus y \leq c \ominus y$, we infer $((x + y) \ominus y) + y$ is defined in $[0, c]$. Then $((x + y) \ominus y) + y = ((x + y) \ominus y) \oplus y = (x \oplus y) \vee y = x \oplus y = x + y$. Property (vi) entails $(x + y) \ominus y = x$. In a similar way we can establish the second property. □

Proposition 4.2. *Let c be a fixed element of a wPEMV-algebra M and let $+$ be the partial addition defined in the last paragraph. Then $([0, c]; \wedge, +_c)$ is a semiclan.*

Proof. For simplicity, we put $+ = +_c$. We have to verify all conditions (S1)–(S5). We note that infima of $x, y \in [0, c]$ taken in M and $[0, c]$ are the same.

(S1) If we put $x = a \oslash b$ and $y = b \ominus a$, we have $a + x = b = y + a$.

(S3) Let $a + x$ and $a + y$ be defined in $[0, c]$. By (iv) of Lemma 4.1, $a + (x \wedge y)$ is defined in $[0, c]$ and by Proposition 3.2(vi), we have $(a+x) \wedge (a+y) = (a \oplus x) \wedge (a \oplus y) = a \oplus (x \wedge y) = x + (a \wedge b)$.

(S4) Let $x+y, (x+y)+z \in [0, c]$. By definition of $+$, we have $z \leq \rho_c(x+y) \leq \rho_c(y)$ which entails that $y + z$ is defined in $[0, c]$. By (W9) and (W4), we conclude

$$z \leq \rho_c(x + y) = (x \oplus y) \oslash c = y \oslash (x \oslash c)$$
$$y + z = y \oplus z \leq y \oplus (y \oslash (x \oslash c)) = y \vee \rho_c(x) = \rho_c(x).$$

So that $x + (y + z)$ is defined in $[0, c]$. Then $x + (y + z) = x \oplus (y \oplus z) = (x \oplus y) \oplus z = (x + y) + z$.

Conversely, let $y+z$ and $x+(y+z)$ be defined in $[0, c]$. Then $x \leq \lambda_c(y+z) \leq \lambda_c(y)$, so that $x + y$ is defined in $[0, c]$. By (W9), $x \leq \lambda_c(y + z) = c \ominus (y + z) = (c \ominus z) \ominus y$

so that $x + y = x \oplus y \leq ((c \ominus z) \ominus y) \oplus y = (c \ominus z) \vee y = \lambda_c(z)$ and $(x+y)+z$ is defined in $[0, c]$. The rest is trivial.

(S2) Let $a + x = a + y$. Then $a + x = a + y \leq c$ and there is $z \in [0, c]$ such that $z + (a + x) = c = z + (a + y)$. By (S4), we have $(z + a) + x = c = (z + b) + y$ and $\lambda_c(\rho_c(z+a)) + x = c = \lambda_c(\rho_c(z+a)) + y$. Lemma 4.1(vi) entails $x = y$. In a similar way we prove that $x + b = y + b$ entails $x = y$.

(S5) Let $(x \wedge y) + z = z$. Then $z = 0 + z$, so by (S2), we have $x \wedge y = 0$. Then by (W4) and (W5), we have $x \vee y = x \oplus (x \ominus y) = x \oplus ((x \wedge y) \ominus y) = x \oplus y \leq c$. In a similar way, $x \vee y = y \oplus x \leq c$. Proposition 3.2(x) yields $(y \oplus x) \ominus x = (y \vee x) \ominus x = (y \ominus x) \vee (x \ominus x) = y \ominus x = y \ominus (x \wedge y) = y$ and $y = (x \vee y) \ominus x \leq c \ominus x = \lambda_c(x)$, so that $y + x$ is defined in $[0, c]$. Similarly, $x + y$ is defined and $x + y = x \oplus y = x \vee y = y \oplus x = y + x$.

Summing up, we conclude that $([0, c]; \wedge, +_c)$ is a semiclan. □

Now, we are ready to present the following important result.

Theorem 4.3. *Let M be a wPEMV-algebra and fix an element $c \in M$. On the interval $[0, c]$, we define the binary operation $\oplus_c$ defined by $(x \oplus_c y) := (x \oplus y) \wedge c$, $x, y \in [0, c]$. Then $([0, c]; \oplus_c, \lambda_c, \rho_c, 0, c)$ is a pseudo MV-algebra.*

In addition, if we put $x \ominus_c y = x \odot_c \lambda_c(y)$ and $x \oslash_c y = \rho_c(x) \odot_c y$, $x, y \in [0, c]$, then the algebra $([0, c]; \vee_c, \wedge_c, \oplus_c, \ominus_c, \oslash_c, 0)$ is a wPEMV-algebra, where $\vee_c$ and $\wedge_c$ are restriction of $\vee$ and $\wedge$ onto $[0, c] \times [0, c]$ and $x \ominus_c y = x \ominus y$ and $x \oslash_c y = x \oslash y$, $x, y \in [0, c]$.

Proof. According to Proposition 4.2, $([0, c]; \wedge, +_c)$ is a semiclan. Due to [5, p. 321], there is an ℓ-group G such that the semiclan $[0, c]$ can be embedded into the semiclan $(G^+; \wedge, +)$ preserving $+, \wedge$ and $\vee$ from $[0, c]$. Let $\theta : [0, c] \to G^+$ be such a mapping. Since $0 \in [0, c]$ and 0 is a neutral element, we have $\theta(0) = 0$ and let $u = \theta(c)$. Define the pseudo MV-algebra $([0, u]; \oplus_u, {}^{-u}, {}^{\sim u}, 0, u)$, where $g \oplus_u h = (g + h) \wedge u$, $g^{-u} = u - g$, $g^{\sim u} = -g + u$, for $g, h \in [0, u]$.

Let $x, y \in [0, c]$. Then $(x \wedge \lambda_c(y)) + y$ is defined in $[0, c]$. We assert that $(x \wedge \lambda_c(y)) + y = (x \oplus y) \wedge c$. Clearly, $(x \wedge \lambda_c(y)) + y \leq x \oplus y, c = \lambda_c(y) + c$. Assume that $z \leq y \leq x \oplus y, \lambda_c(y) + y$. By Proposition 3.2(v), we have $z \ominus y \leq x, \lambda_c(y)$, that is, $z \ominus y \leq x \wedge \lambda_c(y)$ and $z \leq (x \wedge \lambda_c(y)) \oplus y = (x \wedge \lambda_c(y)) + y$ and this proves the assertion.

Therefore, $(x \wedge \lambda_c(y)) + y = (x \oplus y) \wedge c = x \oplus_c y$.

We have $\theta(\lambda_c(x)) = (\theta(x))^{-u}$ and $\theta(\rho_c(x)) = (\theta(x))^{\sim u}$ for each $x \in [0, c]$. Then $\theta(x \oplus_c y) = \theta((x \wedge \lambda_c(y)) + y) = \theta(x \wedge \lambda_c(y)) + \theta(y) = (\theta(x) \wedge \theta(\lambda_c(y))) + \theta(y) = \theta(x) \oplus_u \theta(y)$.

Let $x, y \in [0, c]$. Applying θ to $\rho_c(\lambda_c(y) \oplus_c \lambda_c(x))$ and to $\lambda_c(\rho_c(y) \oplus_c \rho_c(x))$, we see that these elements are the same and they belong to $[0, c]$. Therefore, if we define a binary operation $\odot_c$ on $[0,c]$ by

$$\rho_c(\lambda_c(y) \oplus_c \lambda_c(x)) := x \odot_c y =: \lambda_c(\rho_c(y) \oplus_c \rho_c(x)),$$

then (A1)–(A8) of the definition of pseudo MV-algebras are satisfied. Consequently, the interval $([0,c]; \oplus_c, \lambda_c, \rho_c, 0, c)$ is a pseudo MV-algebra.

Now, check

$$\begin{aligned} x \ominus_c y &= x \odot_c \lambda_c(y) = \lambda_c(y \oplus_c \rho_c(x)) \\ &= c \ominus (y \oplus_c (x \circleddash c)) \\ &= c \ominus (y \oplus (x \circleddash c)) \quad \text{use (W9)} \\ &= (c \ominus (x \circleddash c)) \ominus y = x \ominus y. \end{aligned}$$

In a similar way, we have $x \circleddash_c y = x \circleddash y$. By Example 3.4, the interval $[0,c]$ can be converted into a wPEMV-algebra with top element.

We recall that for each $c \in M$, we have

$$\rho_c(\lambda_c(x) \oplus_c \lambda_c(y)) = \lambda_c(\rho_c(x) \oplus_c \rho_c(y)), \quad x,y \leq c, \tag{4.1}$$

and

$$x \odot_c y = \rho_c(\lambda_c(y) \oplus_c \lambda_c(x)) = \lambda_c(\rho_c(y) \oplus_c \rho_c(x)), \quad x,y \leq c. \tag{4.2}$$

We note that in every pseudo MV-algebra $(A; \oplus, ^-, ^\sim, 0, 1)$ we have $x \leq y$ iff $x^- \oplus y = 1$, see [28, Prop 1.9(a)]. Therefore, if $\leq_c$ is the partial order in the pseudo MV-algebra $([0,c]; \oplus_c, \lambda_c, \rho_c, 0, c)$ forced from (A7), then for $x, y \in [0,c]$, we have $x \leq_c y$ iff $x \leq y$. Therefore, the infima and suprema of elements from $[0,c]$ given by (A7) and (A6), respectively, are the same as original ones in the interval $[0,c]$ or in M. $\square$

In what follows, we define a partial operation $+$ on the whole wPEMV-algebra M and not only on each $[0,c]$.

We define a partial operation $+$ in M as follows: Given $x, y \in M$, $x+y$ is defined in M iff $(x \oplus y) \ominus y = x$, and in such a case, we define $x+y = x \oplus y$. We note that for example, in G^+, this partial operation coincides with the total operation $\oplus$. The basic properties of this partial addition $+$ are gathered in the following lemma.

Lemma 4.4. *Let $+$ be the partial operation defined in the last paragraph on the wPEMV-algebra M. Then for $x, y \in M$, we have*

(i) *The elements $x + 0$ and $0 + x$ are defined in M and $x + 0 = x = 0 + x$.*

(ii) The element $x+y$ is defined in M if and only if $x\oslash(x\oplus y) = y$, equivalently if $(x\oplus y)\ominus y = x$.

(iii) If $x+y$ is defined and $x_1 \leq x$ and $y_1 \leq y$. Then x_1+y_1 is also defined in M.

(iv) If $x+y$ exists in M, then $x+_c y$ exists in $[0,c]$ for each $c \geq x+y$. Moreover, $x+y = x+_c y$.

Proof. (i) It is trivial.

(ii) Let us put $c = x+y$. By Lemma 4.1(i),(vii), we conclude $x \leq \lambda_c(y)$ iff $y \leq \rho_c(x)$. Then $x+y = x+_c y$ which yields $y = x\oslash(x+_c y) = x\oslash(x+y) = x\oslash(x\oplus y)$. In a dual way, we obtain that $x\oslash(x\oplus y)$ implies $(x\oplus y)\ominus x$.

(iii) Set $c = x+y$. Then $x,y,x_1,y_1 \in [0,c]$ and by (iv) and (vii) of Lemma 4.1, we have $x_1+_c y_1$ is defined in $[0,c]$ and $x_1 = (x_1+_c y_1)\ominus y_1 = (x_1\oplus y_1)\ominus y_1$ which means that x_1+y_1 is defined in M.

(iv) Let $c \geq x+y$. Then $x = (x\oplus y)\ominus y \leq c\ominus y$, which entails $x+_c y$ is defined in $[0,c]$ and then $x+y = x\oplus y = x+_c y$. □

We note that (v) in the latter Lemma does not hold for $c \not\geq x+y$, in general. Indeed, let $M = \Gamma(\mathbb{R},1)$ be the MV-algebra of the real interval $[0,1]$. Set $x = 0.5$ and $y = 0.5$. Then $x+y = 1$ is defined in M. Take $c = 0.9$. Then $0.5 = x \not\leq c\ominus y = 0.4$. So that $x+_c y$ does not exist in $[0,c]$.

Proposition 4.5. *The system $(M;\wedge,+)$ is a semiclan.*

Proof. We verify conditions (S1)–(S5) of the definition of the partial addition $+$.

(S1) Let $a \leq b$. Put $c = b$. According to the proof of (S1) in Proposition 4.2, if we set $x = a\oslash b$ and $y = b\ominus a$, then $a+_c x = b = y+_c a$. Applying (vii) of Lemma 4.1, we get that $a+x$ and $y+a$ are defined in M and $a+c = b = y+a$.

(S2) Let $a+x = a+y$. If we set $c = a+x$, then $a+_c x = a+x = a+y = a+_c y$. The proof of (S2) in Proposition 4.2 gives $x = y$. In the same way we establish the second cancelation law of $+$.

(S3) Let $a+x$ and $a+y$ be defined in M. From (iii) of Lemma 4.4, we have $a+(x\wedge y)$ is defined in M. Take $c \geq (a+x)\vee(a+y)$. Then $a+x = a+_c x$, $a+y = a+_c y$ and the proof of (S3) in Proposition 4.2 entails $(a+x)\wedge(a+y) = (a+_c x)\wedge(a+_c y) = a+_c (x\wedge y) = a+(x\wedge y)$.

(S4) Let $x+y$ and $(x+y)+z$ be defined in M. Take $c = (x+y)+z$. Then $x+_c y$ and $(x+_c y)+_c z$ are defined in $[0,c]$. Due to the proof of (S4) in Proposition 4.2, $y+_c z$ and $x+_c(y+_c z)$ are defined in $[0,c]$ and $(x+y)+z = (x+_c y)+_c z = x+_c(y+_c z) = x+(y+z)$ so that $x+(y+z)$ is defined in M. In the same way we can prove the converse.

(S5) If $(x\wedge y)+z = z$, by the cancelation (S2), we have $x\wedge y = 0$. Put $c = x\oplus y$. Then by (S5) of the proof of Proposition 4.5, we have $x+_c y = x\oplus y = x\vee y = y\oplus x = y+_c x$. By Lemma 4.4(ii), we have that $x+y$ and $y+x$ exist in M and $x+y = x\vee y = y+x$.

Summarizing (S1)–(S5), we see that $(M;\wedge,+)$ is a semiclan as stated. □

Theorem 4.6. *Let M be a wPEMV-algebra. Then there is an ℓ-group G such that $(M;\wedge,+)$ can be embedded into $(G^+;\wedge,+)$ preserving all $\wedge$, $\vee$, $+$, $\ominus$ and $\oslash$.*

Proof. Proposition 4.5 asserts that $(M;\wedge,+)$ is a semiclan with the least element 0 that is a neutral element for partial addition $+$. By [5, p. 321], there is an ℓ-group such that $(M;\wedge,+)$ can be embedded into $(G^+;\wedge,+)$ preserving finite meets and finite joins from M. Since it preserves $+$, we have that it preserves also $\ominus$ and $\oslash$. □

If some identity of wPEMV-algebras containing only $\vee, \wedge, \ominus, \oslash$ holds in every positive cone wPEMV-algebra G^+, then due to Theorem 4.6, this identity holds for each wPEMV-algebra. In Section 8 of [24], we will show that it is enough to verify the identity for a special ℓ-group $\mathrm{Aut}(\mathbb{R})$ defined at the beginning of Section 8 of the second part, [24]. For example, we can easily to verify that we have $x\ominus(y\oslash(y\vee x)) = x\wedge y = ((x\vee y)\ominus x)\oslash y$.

Now, we introduce the notions of pseudo effect algebras and of generalized pseudo effect algebras and we show to their relation with wPEMV-algebras.

According to [15, 16], we say that a *pseudo effect algebra* (PEA for short) is a partial algebra $E = (E; +, 0, 1)$, where $+$ is a partial binary operation and 0 and 1 are constants, such that for all $a, b, c \in E$, the following holds

(i) $a+b$ and $(a+b)+c$ exist if and only if $b+c$ and $a+(b+c)$ exist, and in this case $(a+b)+c = a+(b+c)$;

(ii) there is exactly one $d \in E$ and exactly one $e \in E$ such that $a+d = e+a = 1$;

(iii) if $a+b$ exists, there are elements $d, e \in E$ such that $a+b = d+a = b+e$;

(iv) if $1+a$ or $a+1$ exists, then $a = 0$.

If we define $a \leq b$ if and only if there exists an element $c \in E$ such that $a+c = b$, then $\leq$ is a partial ordering on E such that $0 \leq a \leq 1$ for all $a \in E$. It is possible to show that $a \leq b$ if and only if $b = a+c = d+a$ for some $c, d \in E$. Then we write $c = a\,/\,b$ and $d = b\setminus a$, and due to (ii), c and d are uniquely determined. Then

$$(b\setminus a)+a = b = a+(a\,/\,b),$$

and we write $a^- = 1 \setminus a$ and $a^\sim = a / 1$ for all $a \in E$. Then $a^- + a = 1 = a + a^\sim$ and $a^{-\sim} = a = a^{\sim-}$ for all $a \in E$. If the partial operation is commutative, then E is an *effect algebra* in the sense of [25].

A simple corollary of Theorem 4.3 is the following result.

Corollary 4.7. *Let M be a wPEMV-algebra and fix an element $c \in M$. Then the partial algebra $([0, c]; +_c, 0, c)$ is a pseudo effect algebra.*

Proof. Since $([0, c]; \oplus_c, \lambda_c, \rho_c, 0, c)$ is a pseudo MV-algebra, the result is a well-known statement, see e.g. [15, 16]. □

A more general structure than pseudo effect algebras is the class of generalized pseudo effect algebras introduced in [17, 18]. A structure $(E; +, 0)$, where $+$ is a partial binary operation and 0 is a constant, is called a *generalized pseudo effect algebra* (or a GPEA for short) if, for all $a, b, c \in E$, the following hold:

(GP1) $a + b$ and $(a + b) + c$ exist if and only if $b + c$ and $a + (b + c)$ exist, and in this case, $(a + b) + c = a + (b + c)$;

(GP2) if $a + b$ exists, there are elements $d, e \in E$ such that $a + b = d + a = b + e$;

(GP3) if $a + b$ and $a + c$ exist and are equal, then $b = c$; if $b + d$ and $c + d$ exist and are equal, then $b = c$;

(GP4) if $a + b$ exists and $a + b = 0$, then $a = b = 0$;

(GP5) $a + 0$ and $0 + a$ exist and both are equal to a.

A GPEA E is *trivial* if $E = \{0\}$ and it is *non-trivial* if $|E| > 1$. For example, if G^+ is the positive cone of some po-group G, then $(G^+; +, 0)$ is a GPEA.

In the same way as for pseudo effect algebras, we introduce a binary relation $\leq$ in a GPEA E: For $a, b \in E$, we define $a \leq b$ if and only if there is an element $c \in E$ such that $a + c = b$. Equivalently, there exists an element $d \in E$ such that $d + a = b$. Then $\leq$ is a partial order on E (it is the so-called GPEA-order); sometimes we write $x \leq_E y$ to underline that the order is taken in the GPEA E.

If the partial operation $+$ on E is commutative, then a GPEA E is said to be a *generalized effect algebra*, GEA for short. A GPEA E is *weakly commutative* if $a + b$ exists in E iff $b + a$ is defined in E.

We introduce also two partial binary operations \ and / on a GPEA E in the same way as for pseudo effect algebras: For any $a, b \in E$, a / b is defined if and only if $b \setminus a$ is defined if and only if $a \leq b$, and in such a case we have $(b \setminus a) + a = b = a + (a / b)$. Then $a = (b \setminus a) / b = b \setminus (a / b)$.

For example, if G is a po-group, then the positive cone G^+ with the group addition $+$ and 0 gives an example of a GPEA.

We recall that a non-empty set I of a PEA E is an *ideal* (more precisely a PEA-ideal) of E if (i) if $x \in E$, $y \in I$, and $x \leq y$, then $x \in I$, and (ii) if $x, y \in I$ and $x+y$ is defined in E, then $x+y \in I$. An ideal I is (i) *normal* if $x + I = I + x$ for each $x \in E$, where $x + I = \{x + i \colon i \in I, x + i \in E\}$ and dually we define $I + x$; (ii) *maximal* if it is a proper subset of E not contained properly in any proper ideal of E.

Now we introduce two kinds of the Riesz Decomposition Property (RDP for short), for more details, see [15, 16]. We say that a GPEA E satisfies

(i) RDP if, for all $a_1, a_2, b_1, b_2 \in E$ such that $a_1 + a_2 = b_1 + b_2$, there are four elements $c_{11}, c_{12}, c_{21}, c_{22} \in E$ such that $a_1 = c_{11} + c_{12}$, $a_2 = c_{21} + c_{22}$, $b_1 = c_{11} + c_{21}$ and $b_2 = c_{12} + c_{22}$; this property will be formally denoted by the following RDP table:

a_1	c_{11}	c_{12}
a_2	c_{21}	c_{22}
	b_1	b_2

;

(ii) RDP$_2$ if, for all $a_1, a_2, b_1, b_2 \in E$ such that $a_1 + a_2 = b_1 + b_2$, there are four elements $c_{11}, c_{12}, c_{21}, c_{22} \in E$ such that $a_1 = c_{11} + c_{12}$, $a_2 = c_{21} + c_{22}$, $b_1 = c_{11} + c_{21}$ and $b_2 = c_{12} + c_{22}$, and $c_{12} \wedge c_{21} = 0$; it can be described by the so-called RDP$_2$ table which is an RDP table with $c_{12} \wedge c_{21} = 0$. In such the case, the table is unique. In other words RDP$_2$ is RDP + $c_{12} \wedge c_{21} = 0$.

We note that RDP$_2$ implies RDP but the converse is not guaranteed. It holds, if E is commutative. Moreover, if a PEA satisfies RDP$_2$, then it is lattice ordered and equivalent to a pseudo MV-algebra, see [16].

Theorem 4.8. *Let M be a wPEMV-algebra. Then the partial algebra $(M; +, 0)$ is a GPEA with* RDP$_2$, *where the order on M induced by $\vee$ and $\wedge$ coincides with the GPEA-order.*

Moreover, if $x, y \in M$, then

$$x \ominus y = x \setminus (x \wedge y), \quad y \oslash x = (x \wedge y) / x. \qquad (4.3)$$

Proof. Since $(M; \wedge, 0)$ is a semiclan, see Theorem 4.6, then (GP1), (GP2), (GP3) and (GP5) hold and since $a, b \leq a+b$, (GP4) holds, too. Therefore, $(M; +, 0)$ is a GPEA. Let $a_1 + a_2 = b_1 + b_2$. Put $c = a_1 + a_2$. Due to Theorem 4.3, $([0, c]; \oplus_c, \lambda_c, \rho_c, 0, c)$

is a pseudo MV-algebra and therefore, for it RDP_2 holds, see e.g. [14, Prop 5.1], that is, there are four elements $c_{11}, c_{12}, c_{21}, c_{22} \in [0, c]$ such that $a_1 = c_{11} +_c c_{12}$, $a_2 = c_{21} +_c c_{22}$, $b_1 = c_{11} +_c c_{21}$, and $b_2 = c_{12} +_c c_{22}$. Lemma 4.4(v) implies that in the last four equalities we can change $+_c$ by $+$.

Let $\leq$ be the order generated by the lattice operations $\vee$ and $\wedge$ defined in the wPEMV-algebra M. We set $x \preceq y$ iff there is $z \in M$ such that $x + z = y$. Take an element $a \in M$ such that $x \oplus y \oplus z \leq a$. Then $x + z = x \oplus z = y$ which means also $x \leq y$. Now let $x \leq y$, then $y = x \vee y = x \oplus (\rho_a(x) \odot y) = x + (\rho_a(x) \odot y)$, i.e. $x \preceq y$ and $\preceq = \leq$. Equalities in (4.3) follow from easy calculations in the pseudo MV-algebra $([0, a]; \oplus, \lambda_a, \rho_a, 0, a)$. $\square$

In the rest of the chapter, we show a close relationship of wPEMV-algebras with integral GMV-algebras, a special class of residuated lattices.

We recall that due to [2, 4, 27], a *residuated lattice* is an algebra $(L; \wedge, \vee, \cdot, \backslash, /, e)$ such that $(L; \wedge, \vee)$ is a lattice, $(L; \cdot, e)$ is a monoid, and for all $x, y, z \in L$,

$$x \cdot y \leq z \quad \Leftrightarrow \quad x \leq z/y \quad \Leftrightarrow \quad y \leq x \backslash z.$$

If no confusion, we write also $xy = x \cdot y$. For example, if G is an ℓ-group written in a multiplicative way with unit e, then $(G; \wedge, \vee, \cdot, ^{-1}, \backslash, /, e)$ is a residuated lattice with $x/y = xy^{-1}$ and $x \backslash y = x^{-1} y$. Similarly, the negative cone $(G^-; \wedge, \vee, \cdot, ^{-1}, \backslash, /, e)$ with $x/y = (xy^{-1}) \wedge e$ and $x \backslash y = (x^{-1} y) \wedge e$ is also a residuated lattice. Residuated lattices form a finitely based equational class of algebras.

A residuated lattice L is (i) *integral* if $x \leq e$ for each $x \in L$, (ii) a *generalized MV-algebra* (GMV-algebra, for short) if it satisfies the identities

$$x/((x \vee y) \backslash x) = x \vee y = (x/(x \vee y)) \backslash x,$$

and (iii) *bounded* if it is integral and there is an element $a \in L$ such that $a \leq x$ for each $x \in L$.

Proposition 4.9. *Every wPEMV-algebra is equivalent to an integral GMV-algebra.*

Proof. Let $(M; \vee, \wedge, \oplus, \ominus, \oslash, 0)$ be a wPEMV-algebra. Let $\leq$ be the partial order defined on M. On M, we define the reverse order $\preceq$ to the order $\leq$ by $x \preceq y$ iff $y \leq x$, and we determine the following binary operations: $x \cdot y = x \oplus y$, $x/y = x \ominus y$, $x \backslash y = x \oslash y$, $x \sqcap y = x \vee y$, and $x \sqcup y = x \wedge y$. Then an easy calculation shows that $(M; \sqcap, \sqcup, \cdot, \backslash, /, 0)$ is an integral GMV-algebra.

Conversely, let $(M; \sqcap, \sqcup, \cdot, \backslash, /, 0)$ be an integral GMV-algebra with the partial order $\preceq$.

On the set M, we define $x \oplus y = x \cdot y$, $x \ominus y = x/y$, $x \obslash y = x\backslash y$, $x \vee y = x \sqcap y$ and $x \wedge y = x \sqcup y$. Let $\leq$ be the reverse order to the order $\preceq$ determined by $\sqcap$ and $\sqcup$. In what follows, we show that $(M; \vee, \wedge, \oplus, \ominus, \obslash, 0)$ is a wPEMV-algebra. We exhibit conditions (W1)–(W10) of the definition of wPEMV-algebras, see 3.1.

(W1) By [27, Lem 2.9], $(M; \wedge, \vee)$ is a distributive lattice.

(W2) By definition, $(M; \cdot, 0)$ is a monoid.

(W3) From [4, Lem 3.2(5)], it follows that $(y \oplus x) \ominus x = (y \cdot x)/x \succeq y \cdot (x/x) = y \cdot 0 = y$ (by [27, Lem 2.7(ii)]).

(W4) By [27, Lem 2.11(i)], $(y \ominus x) \oplus x = (y/x) \cdot x$ is the the greatest lower bound of x and y in the integral GMV-algebra M. So, $(y \ominus x) \oplus x = x \vee y$ in the poset $(M; \leq)$.

(W5) By [4, Lem 3.2(3)], $x \ominus (x \wedge y) = x/(x \vee y) = (x/x) \wedge (x/y) = 0 \wedge (x/y) = x/y = x \ominus y$.

(W6) $y \ominus (x \obslash y) = y/(x\backslash y)$ is the least upper bound for x and y in the integral GMV-algebra M. So, $y \ominus (x \obslash y)$ is $x \wedge y$ in the poset $(M; \leq)$.

(W7) By [27, Lem 2.13(iii)], $z \ominus (x \vee y) = z/(x \wedge y) = (z/x) \vee (z/y)$. So, $z \ominus (x \vee y) = (z \ominus x) \wedge (z \ominus y)$.

(W8) It is similar to (W7), and it follows from [27, Lem 2.13(iii)].

(W9) By [4, Lem 3.2(10)], $x \ominus (y \oplus z) = x/(y \cdot z) = (x/z)/y = (x \ominus z) \ominus y$.

(W10) It simply follows from [4, Lemma 3.2(1.a)]. □

From this proposition and [27, Thm B], we conclude that every wPEMV-algebra can be embedded into a negative cone G^- of some ℓ-group G, see also Theorem 4.6. In general, due to Proposition 4.9, our results can reformulated in the language of integral GMV-algebras. Therefore, every bounded integral GMV-algebra is in fact a pseudo MV-algebra, compare with [27, p. 277]. The main difference between wPEMV-algebras and integral GMV-algebras is that the first ones live in the positive cone and the second ones do in the negative cone of some ℓ-groups.

5 Congruences, Ideals and Representing wPEMV-algebras

We continue in presenting basic properties of wPEMV-algebras and we introduce representing wPEMV-algebras corresponding to wPEMV-algebras where top element is not assumed a priori.

As a direct corollary of Theorem 4.3 we present the following result.

Corollary 5.1. *If $(M; \vee, \wedge, \oplus, \ominus, \oslash, 0)$ is a wPEMV-algebra with top element 1, then the algebra $(M; \oplus, ^-, ^\sim, 0, 1)$ is a pseudo MV-algebra, where $x^- = 1 \ominus x$ and $x^\sim = x \oslash 1$, for all $x \in M$.*

Proposition 5.2. *For every $a \in M$, we have*
$$(x \oplus y) \wedge a = (x \wedge a) \oplus_a (y \wedge a), \quad x, y \in M.$$

Proof. Let $x, y, a \in M$. Choose $b \in M$ such that $x \oplus y \oplus a \leq b$. Due to Theorem 4.3, $([0, b]; \oplus, \lambda_b, \rho_b, 0, b)$ is a pseudo MV-algebra. By [28, Prop 1.17], $(x \oplus_b y) \wedge a \leq (x \wedge a) \oplus_b (y \wedge a)$, which implies that $(x \oplus y) \wedge a = (x \oplus_b y) \wedge a \leq (x \wedge a) \oplus_b (y \wedge a)$ and so $(x \oplus y) \wedge a \leq ((x \wedge a) \oplus_b (y \wedge a)) \wedge a = (x \wedge a) \oplus_a (y \wedge a)$. On the other hand, $x \wedge a \leq x$ and $y \wedge a \leq y$ and so by Proposition 3.2(i), $(x \wedge a) \oplus_a (y \wedge a) \leq (x \wedge a) \oplus (y \wedge a) \leq x \oplus y$. It follows that $(x \wedge a) \oplus_a (y \wedge a) \leq (x \oplus y) \wedge a$. That is, $(x \wedge a) \oplus_a (y \wedge a) = (x \oplus y) \wedge a$. □

For wPEMV-algebras, we define $\lambda_a(x) = a \ominus x$ and $\rho_a(x) = x \oslash a$, for $x \leq a$. Now, we present a useful proposition concerning functions λ_a and ρ_a if a is an idempotent.

Proposition 5.3. *Let $(M; \vee, \wedge, \oplus, \ominus, \oslash, 0)$ be a wPEMV-algebra, $a, b \in \mathcal{I}(M)$ such that $a \leq b$. Then, for each $x \in [0, a]$, we have:*

(i) $\lambda_b(a) = \rho_b(a)$ *is an idempotent.*

(ii) $\lambda_a(x) = \lambda_b(x) \wedge a$ *and* $\rho_a(x) = \rho_b(x) \wedge a$.

(iii) $\lambda_b(x) = \lambda_a(x) \oplus \lambda_b(a) = \lambda_b(a) \oplus \lambda_a(x)$ *and* $\rho_b(x) = \rho_a(x) \oplus \rho_b(a) = \rho_b(a) \oplus \rho_a(x)$.

(iv) $\rho_a(\lambda_a(x)) = x = \lambda_a(\rho_a(x))$.

(v) $\lambda_a(x) \leq \lambda_b(x)$ *and* $\rho_a(x) \leq \rho_b(x)$.

(vi) *Let a, b be arbitrary elements of M such that $a \leq b$. For $x \leq a$, we have* $\lambda_b(x) = \lambda_b(a) \oplus \lambda_a(x)$ *and* $\rho_b(x) = \rho_a(x) \oplus \rho_b(a)$.

(vii) *Let a, b be arbitrary elements of M such that $a \leq b$. For $x \leq a$, we have* $\lambda_b(x) \wedge a = (\lambda_b(a) \wedge a) \oplus_a \lambda_a(x)$ *and* $\rho_b(x) \wedge a = \rho_a(x) \oplus_a (\rho_b(a) \wedge a)$.

Proof. According to Theorem 4.3, $([0, a]; \oplus_a, \lambda_a, \rho_a, 0, a)$ is a pseudo MV-algebra and $x \oplus_a y = x \oplus y$ for all $x, y \in [0, a]$. The same is true for $([0, b]; \oplus_b, \lambda_b, \rho_b, 0, b)$.

(i) By Proposition 3.2(xiii) and [28, Prop 4.3(1)], the elements $\lambda_b(a)$ and $\rho_b(a)$ are Boolean elements and they coincide.

(ii) We use the both-side distributivity of $\oplus$ with respect to join and meet in pseudo MV-algebras, see [28, Prop 1.15, 1.21]: From $\lambda_b(x) \wedge a \in [0, a] \subseteq [0, b]$ and

$(\lambda_b(x) \wedge a) \oplus x = (\lambda_b(x) \oplus x) \wedge (a \oplus x) = b \wedge (a \vee x) = a$, we have $\lambda_a(x) \leq \lambda_b(x) \wedge a$. Whence, $b = \lambda_b(a) \oplus a = (\lambda_b(a) \oplus \lambda_a(x)) \oplus x$ yielding $\lambda_b(a) \oplus \lambda_a(x) \geq \lambda_b(x)$. Therefore, $(\lambda_b(a) \oplus \lambda_a(x)) \wedge a \geq \lambda_b(x) \wedge a$. Since $\lambda_b(a)$ is a Boolean element of M, we have $\lambda_b(x) \wedge a \leq (\lambda_b(a) \oplus \lambda_a(x)) \wedge a = (\lambda_b(a) \vee \lambda_a(x)) \wedge a = \lambda_a(x) \wedge a = \lambda_a(x)$. Finally, we get $\lambda_a(x) = \lambda_b(x) \wedge a$.

In a dual way we establish $\rho_a(x) = \rho_b(x) \wedge a$.

To prove (iii)–(v), we follow ideas from [19, Prop 3.2].

(vi) Since $(b \ominus a) \oplus (a \ominus x) \leq (b \ominus a) \oplus a = b$, see (W4), we have due to Theorem 4.3:
$$\rho_b(\lambda_b(a) \oplus \lambda_a(x)) = \rho_b(\lambda_b(a) \oplus_b \lambda_a(x)) =$$
$$\rho_b(\lambda_a(x)) \odot_b a = \lambda_a(x) \oslash_b a = \lambda_a(x) \oslash a = (a \ominus x) \oslash a = x,$$

see Proposition 3.2(ii), which establishes the first equality. The second one can be proved analogously.

(vii) It follows from (vi) and Proposition 5.2. □

Let $(M_1; \vee, \wedge, \oplus, \ominus, \oslash, 0)$ and $(M_2; \vee, \wedge, \oplus, \ominus, \oslash, 0)$ be two wPEMV-algebras. We know that a map $f : M_1 \to M_2$ is a *homomorphism* if f preserves all operations $\vee, \wedge, \oplus, \ominus, \oslash$ and 0. That is, $f(x * y) = f(x) * f(y)$ for all $x, y \in M_1$ and all $* \in \{\vee, \wedge, \oplus, \ominus, \oslash\}$ and $f(0) = 0$. A one-to-one homomorphism $f : M_1 \to M_2$ is called an *embedding*. If there is an embedding from M_1 to M_2, then we say that M_1 can be embedded in M_2.

A non-empty set I of a wPEMV-algebra $(M; \vee, \wedge, \oplus, \ominus, \oslash, 0)$ is called an *ideal* if I is closed under $\oplus$ and for each $x, y \in M$, $x \leq y \in I$ implies $x \in I$. Since $x \ominus y \leq x$ and $x \oslash y \leq y$ (see Proposition 3.2(xi)), then clearly each ideal I is closed under $\oslash$, $\ominus$ and $\odot$, too.

If A is a subset of M, then the ideal $I_0(A)$ generated by A is the set $I_0(A) = \{x \in M \mid x \leq a_1 \oplus \cdots \oplus a_n, a_1, \ldots, a_n \in A\}$ if A is non-empty, otherwise, $I_0(A) = \{0\}$. In particular, if $A = \{a\}$, we write simply $I_0(a) := I_0(\{a\})$. We note that the following property

$$x \wedge (y_1 \oplus y_2) \leq (x \wedge y_1) \oplus (x \wedge y_2) \tag{5.1}$$

holds in every pseudo MV-algebra, [28, Prop 1.17], so that it holds also in every wPEMV-algebra in view of Theorem 4.3. Therefore, it is possible to show that the system of all ideals of M is a distributive lattice. Moreover, if $x, y \in M$, then by (5.1), we have

$$I_0(x) \wedge I_0(y) = I_0(x \wedge y).$$

We recall that an ideal I of a wPEMV-algebra M is *prime* if, for each $x, y \in M$, $x \wedge y \in I$ implies that $x \in I$ or $y \in I$.

Lemma 5.4. *Let I be an ideal of a wPEMV-algebra M. The following statements are equivalent:*

(i) *I is prime.*

(ii) *If $x, y \in M \setminus I$, then $x \wedge y > 0$.*

(iii) *If $x \wedge y = 0$, then $x \in I$ or $y \in I$.*

(iv) *For all $x, y \in M$, $x \ominus y \in I$ or $y \ominus x \in I$.*

(v) *For all $x, y \in M$, $x \oslash y \in M$ or $y \oslash x \in I$.*

(vi) *If A and B are ideals of M such that $I \subseteq A \cap B$, then $A \subseteq B$ or $B \subseteq A$.*

(vii) *If A and B are ideals of M such that $I \subsetneq A$ and $I \subsetneq B$, then $I \subsetneq A \cap B$.*

As a corollary, if I is prime and A is an ideal of M such that $I \subseteq A$, then A is prime as well.

Proof. (i) $\Rightarrow$ (ii). Assume the converse, i.e. $x \wedge y = 0$ for some $x, y \in M \setminus I$. Since $x \wedge y \in I$, we have $x \in I$ or $y \in I$, a contradiction.

(ii) $\Rightarrow$ (i). Assume $x \wedge y \in I$. Due to Proposition 3.2(xii), we have $(x \ominus y) \wedge (y \ominus x) = 0$, so that $x \ominus y \in I$ or $y \ominus x \in I$ (otherwise $x \ominus y, y \ominus x \notin I$, a contradiction with $(x \ominus y) \wedge (y \ominus x) = 0$). Suppose $x \ominus y \in I$. Then $x = (x \ominus y) \oplus (x \wedge y) \in I$; analogously in the second case. Whence, I is prime.

The implication (i) $\Rightarrow$ (iii) is trivial as well as the implication (iii) $\Rightarrow$ (ii).

(i) $\Leftrightarrow$ (iv), (v). They follow from Proposition 3.2(xii).

(i) $\Rightarrow$ (vi). Let $I \subseteq A, B$. If A and B are not comparable, there are $x \in A \setminus B$ and $y \in B \setminus A$. Since $(x \ominus y) \wedge (y \ominus x) = 0$, then by (i) and (iv), $x \ominus y \in I$ or $y \ominus x \in I$. Assume, say, $x \ominus y \in I$. Due to (W4), we have $(x \ominus y) \oplus y = x \vee y = (y \ominus x) \oplus x$ which implies $x \vee y \in B$, a contradiction because $x \notin B$.

(vi) $\Rightarrow$ (i). Let $x \wedge y \in I$. Then $I \subseteq (I \vee I_0(x)), (I \vee I_0(y))$. We can assume that $I \vee I_0(x) \subseteq I \vee I_0(y)$. Then $(I \vee I_0(x)) \wedge (I \vee I_0(y)) = I \vee (I_0(x) \wedge I_0(y)) = I \vee I_0(x \wedge y) = I$ which implies $x \in I$.

(i) $\Rightarrow$ (vii). Assume the converse, i.e. $I = A \cap B$. Property (vi) shows that $A \subseteq B$ or $B \subseteq A$ which yields $I = A \cap B = A$ or $I = A \cap B = B$, a contradiction.

(vii) $\Rightarrow$ (i). Let $x \wedge y \in I$ and let $x, y \notin I$. Then $I \subsetneq I \vee I_0(x)$ and $I \subsetneq I \vee I_0(y)$, so that $I \subsetneq (I \vee I_0(x)) \cap (I \vee I_0(y)) = I \vee I_0(x \wedge y)$ which entails $x \wedge y \notin I$, a contradiction.

The last corollary follows from (iii). $\square$

We note that in Corollary 6.9, it will be shown that the set of normal ideals of a wPEMV-algebra is a distributive lattice with respect to the set-theoretical inclusion.

Lemma 5.5. *Let $(M; \vee, \wedge, \oplus, \ominus, \otimes, 0)$ be a wPEMV-algebra. Then, for each $x, y, z \in M$, we have*

(i) $x \ominus z \leq (x \ominus y) \oplus (y \ominus z)$.

(ii) $x \otimes z \leq (x \otimes y) \oplus (y \otimes z)$.

(iii) $(x \vee y) \ominus y = x \ominus y$ *and* $y \otimes (x \vee y) = y \otimes x$.

Proof. (i) Let $x, y, z \in M$. By (W9), $(x \ominus z) \ominus (y \ominus z) = x \ominus ((y \ominus z) \oplus z) = x \ominus (y \vee z) \leq x \ominus y$. Proposition 3.2(v) implies that $x \ominus z \leq (x \ominus y) \oplus (y \ominus z)$.

(ii) It is similar to (i).

(iii) Let $x, y \in M$. By (W4) and (W3), $(x \vee y) \ominus y = ((x \ominus y) \oplus y) \ominus y \leq x \ominus y$. On the other hand, by Proposition 3.2(iv), $(x \vee y) \ominus y \geq x \ominus y$. Hence $(x \vee y) \ominus y = x \ominus y$. In a similar way, we can prove the second equality. $\square$

An ideal I of a wPEMV-algebra $(M; \vee, \wedge, \oplus, \ominus, \otimes, 0)$ is *normal* if, for each $x, y \in M$, $y \ominus x \in I$ if and only if $x \otimes y \in I$. Equivalently, $x \oplus I = I \oplus x$ for each $x \in M$. Indeed, I be normal and let $x \ominus y \in I$. Then there is $z \in I$ such that $x \vee y = (x \ominus y) \oplus y = y \oplus z$ which gives $y \otimes x = y \otimes (x \vee y) = y \otimes (y \oplus z) \leq z$, so that $y \otimes x \in I$. Conversely, let $x \in M$ and $y \in I$. Then $x \otimes (x \oplus y) \leq y \in I$, so that $y_1 := (x \oplus y) \ominus x \in I$. Therefore, $x \oplus y = y_1 \oplus x$.

Let I be a normal ideal of M. Then the relation Θ_I defined by

$$\Theta_I := \{(x, y) \in M \times M \mid x \ominus y, x \otimes y \in I\} \tag{5.2}$$

is an equivalence relation on the wPEMV-algebra $(M; \vee, \wedge, \oplus, \ominus, \otimes, 0)$. Clearly, Θ_I is reflexive. Since I is normal, we can easily see that Θ_I is symmetric. Let $(x, y), (y, z) \in \Theta_I$. Then $x \ominus y, x \otimes y \in I$ and so $y \otimes x, y \ominus x \in I$, too. Similarly, $y \ominus z, y \otimes z \in I$ and $z \otimes y, z \ominus y \in I$. From Lemma 5.5 we get that $x \ominus z \leq (x \ominus y) \oplus (y \ominus z) \in I$ and $x \otimes z \leq (x \otimes y) \oplus (y \otimes z) \in I$. Consequently, Θ_I is transitive.

For example, let $x \in M$. If $(x, 0) \in \Theta_I$, then $x = x \ominus 0 \in I$. Conversely, if $x \in I$, then $x \ominus 0 = x \in I$ (by Proposition 3.2(viii)) and $x \otimes 0 = 0 \in I$ (by Proposition 3.2(ix)). Hence

$$(x, 0) \in \Theta_I \Leftrightarrow x \in I. \tag{5.3}$$

Proposition 5.6. *Let I be a normal ideal of a wPEMV-algebra $(M; \vee, \wedge, \oplus, \ominus, \otimes, 0)$. Then Θ_I is a congruence relation on M.*

Conversely, if Θ is a congruence of M, then $I_\Theta := \{x \in M \mid (x, 0) \in \Theta\}$ is a normal ideal of M such that $\Theta = \Theta_{I_\Theta}$. There is a one-to-one correspondence between normal ideals and congruences given by $\Theta \leftrightarrow I_\Theta$ and $I \leftrightarrow I_{\Theta_I}$. Moreover, $\Theta_I \subseteq \Theta_J$ if and only if $I \subseteq J$.

Proof. Since Θ_I is an equivalence relation, it suffices to show that, for all $(x, y) \in \Theta_I$ and all $a \in M$, we have $(a * x, a * y), (x * a, y * a) \in \Theta_I$ for all $* \in \{\vee, \wedge, \oplus, \odot, \ominus\}$. Let $(x, y) \in \Theta_I$ and $a \in M$. Then $x \ominus y, x \odot y \in I$ and so $y \odot x, y \ominus x \in I$, since I is normal.

(1) By (W9) and Proposition 3.2(v), $(x \oplus a) \ominus (y \oplus a) = ((x \oplus a) \ominus a) \ominus y \leq x \ominus y \in I$. Similarly, $(y \oplus a) \ominus (x \oplus a) \leq y \ominus x \in I$. Since I is normal, we get that $(x \oplus a, y \oplus a) \in \Theta_I$. Also, $(a \oplus x) \odot (a \oplus y) = x \odot (a \odot (a \oplus y)) \leq x \odot y \in I$, by (W9) and Proposition 3.2(v). Thus, $(a \oplus x) \odot (a \oplus y), (a \oplus y) \odot (a \oplus x) \in I$. Now, from normality of I we get that $(a \oplus x) \ominus (a \oplus y), (a \oplus y) \ominus (a \oplus x) \in I$. Therefore, $(a \oplus x, a \oplus y) \in \Theta_I$.

(2) By (W4) and Proposition 3.2(v), $(x \ominus a) \ominus (y \ominus a) = x \ominus ((y \ominus a) \oplus a) = x \ominus (y \vee a) \leq x \ominus y \in I$. Analogously, $(y \ominus a) \ominus (x \ominus a) \in I$. Now, since I is normal, we get $(x \ominus a) \odot (y \ominus a) \in I$ which implies that $(x \ominus a, y \ominus a) \in \Theta_I$. Also, by (W4) and Proposition 3.2(v), $(a \odot x) \odot (a \odot y) = (a \oplus (a \odot x)) \odot y = (a \vee x) \odot y \leq x \odot y \in I$. Similarly, $(a \odot y) \odot (a \odot x) \leq y \odot x \in I$ and so $(a \odot x) \ominus (a \odot y) \in I$ (since I is normal). Therefore, $(a \odot x, a \odot y) \in \Theta_I$.

(3) From (W4) and (W6) we can simply deduce that $(a \vee x, a \vee y), (a \wedge x, a \wedge y) \in \Theta_I$.

From (1), (2) and (3) it follows that Θ_I is a congruence relation on M.

If now Θ is a congruence on M, then it is easy to see that $I_\Theta = \{x \in M \mid (x, 0) \in \Theta\}$ is a normal ideal and $\Theta = \Theta_{I_\Theta}$.

Finally, the equivalence $\Theta_I \subseteq \Theta_J$ iff $I \subseteq J$ for normal ideals I and J is now evident. $\square$

Corollary 5.7. *If I is a normal ideal of a wPEMV-algebra $(M; \vee, \wedge, \oplus, \ominus, \odot, 0)$, then for each $x \in M$, the equivalence class of Θ_I containing x is denoted by x/Θ_I or simply x/I. Also, M/I will denote the set $\{x/I \mid x \in M\}$. Since I is a normal ideal, by Proposition 5.6, Θ_I is a congruence relation and so M/I with the following binary operations and the nullary operation $0/I$ is a wPEMV-algebra.*

$$x/I * y/I = (x * y)/I, \quad \forall * \in \{\vee, \wedge, \oplus, \odot, \ominus\}. \tag{5.4}$$

Proof. It is straightforward by Proposition 5.6. $\square$

A wPEMV-algebra $(M; \vee, \wedge, \oplus, \ominus, \odot, 0)$ is said to be (i) *strict* if for each $x \in M$, $x \oplus x = x$ implies that $x = 0$, and (ii) *cancellative* if $a \oplus b_1 = a \oplus b_2$ and $a_1 \oplus b = a_2 \oplus b$ iff $b_1 = b_1$ and $a_1 = a_2$.

Theorem 5.8. (1) *A linearly ordered wPEMV-algebra* $(M; \vee, \wedge, \oplus, \ominus, \oslash, 0)$ *is strict or it has a top element.*

(2) *Let* $(M; \vee, \wedge, \oplus, \ominus, 0)$ *be a cancellative wPEMV-algebra. Then it is isomorphic to the wPEMV-algebra of a positive cone of some ℓ-group* $(G; +, 0)$.

(3) *A linearly ordered wPEMV-algebra* $(M; \vee, \wedge, \oplus, \ominus, 0)$ *is strict if and only if M is cancellative.*

Proof. (1) Suppose that M is not strict. Then there exists $a \in M \setminus \{0\}$ such that $a \oplus a = a$. By Theorem 4.3, we know that $([0, a]; \oplus_a, \lambda_a, \rho_a, 0, a)$ is a pseudo MV-algebra and $([0, a]; \vee, \wedge, \oplus, \ominus, \oslash, 0)$ is a wPEMV-algebra with top element. Clearly, $x \oplus_a y = x \oplus y$ for each $x, y \in [0, a]$ (since a is idempotent). We claim that $M = [0, a]$. Let x be an arbitrary element of M. Set $b = x \oplus a \oplus x$. Then $a \leq b$. Consider the pseudo MV-algebra $([0, b]; \oplus_b, \lambda_b, \rho_b, 0, b)$ which is a chain. In this linearly ordered pseudo MV-algebra we have $a \oplus_b a = (a \oplus a) \wedge b = a \wedge b = a$ and $b \oplus_b b = b$, that is a and b are Boolean elements in the pseudo MV-algebra $[0, b]$ such that $0 < a \leq b$. We assert $a = b$. If not then $\lambda_b(a) = \rho_b(a) < b$ are also Boolean elements in the pseudo MV-algebra $[0, b]$. There are two cases: (i) $a \leq \lambda_b(a)$ or (ii) $\lambda_b(a) < a$. In the first case, Proposition 3.2(xiii) implies $b = \lambda_b(a) \oplus_a a = \lambda_b(a) \vee a = \lambda_b(a) < b$ and in the second case, $b = \lambda_b(a) \oplus_a a = \lambda_b(a) \vee a = a < b$, both are contradictions. So that $a = b$, and in particular $x \leq a$. Therefore, $M = [0, a]$.

(2) Let M be a cancellative wPEMV-algebra. Due to Proposition 3.2(i), M is both right and left naturally ordered. Applying the Nakada theorem, [26, Prop X.1], there is an ℓ-group G (not necessarily Abelian) such that M is isomorphic to $(G^+; \vee, \wedge, \oplus, \ominus, \oslash, 0)$, see Example 3.3.

(3) Let $(M; \vee, \wedge, \oplus, \ominus, \oslash, 0)$ be a linearly ordered strict wPEMV-algebra and $x, y, z \in M$ be such that $x \oplus z = y \oplus z$. If $z = 0$, then trivially $x = y$. Assume thus $z > 0$. Take $b \in M$ such that $2.(x \oplus z \oplus y \oplus z) \leq b$. Consider the pseudo MV-algebra $([0, b]; \oplus, \lambda_b, \rho_b, 0, b)$ (by Theorem 4.3). From [14, Thm 3.6] it follows that there exists a unital linearly ordered group (G, u) such that $\Gamma(G, u)$ and $[0, b]$ are isomorphic pseudo MV-algebras. Without loss of generality we can suppose that $[0, b] = \Gamma(G, u)$ and so $u = b$. In the pseudo MV-algebra $[0, b]$, we have $x \oplus_b z = (x \oplus z) \wedge b$. By the assumptions, $x \oplus z \lneq b$ (otherwise $x \oplus z = b$ is an idempotent and the strictness of M yields $x \oplus z = 0$ and $x = 0 = z$, an absurd), so that $x \oplus_b z = x \oplus z$. On the other hand, in the pseudo MV-algebra $\Gamma(G, u)$ we have $x \oplus_b z = (x + z) \wedge b$, where $+$ is the group addition the group G. Hence, $x + z \lneq b$ (otherwise, since G is linearly ordered, $b \leq x + z$ means that $x \oplus_b z = (x + z) \wedge b = b$, which is absurd) and so $x \oplus z = x \oplus_b z = x + z$. A similar argument shows that $y \oplus z = y + z$. Now, $x + z = y + z$ and hence $x = y$. In a similar way, we can show that $z \oplus x = z \oplus y$ implies that $x = y$. Therefore, $(M; \vee, \wedge, \oplus, \ominus, 0)$ is cancellative.

Conversely, let M be cancellative and let $x \oplus x = x$. Then $x \oplus 0 = x \oplus x$, so that $x = 0$, and M is strict. □

Using the language of integral GMV-algebras, (2) was established in [2].

We note that a linearly ordered wPEMV-algebra $M \neq \{0\}$ cannot be simultaneously strict and with top element. Moreover, every cancellative wPEMV-algebra is strict.

Proposition 5.9. *Let $(M; \vee, \wedge, \oplus, \ominus, \obar, 0)$ be a wPEMV-algebra and $I_1 := \downarrow \mathcal{I}(M)$. We assert I_1 is a normal ideal of M and M/I_1 is a strict wPEMV-algebra. Similarly, $I_2 = \{x \in M \mid x \wedge a = 0, \; \forall a \in \mathcal{I}(M)\}$ is an ideal and a strict subalgebra of M.*

Proof. First, we show that I_1 is a normal ideal of M. Indeed, (i) clearly I_1 is an ideal of M. (ii) If $x, y \in M$ such that $x \ominus y \leq a \in \mathcal{I}(M)$, then by Proposition 3.2(v), $x \leq a \oplus y = y \oplus a$ and similarly, $y \obar x \leq a$. That is, M_1 is normal.

Consider the wPEMV-algebra M/I_1. We claim that it is strict. Let $x \in M$ be such that x/I_1 is an idempotent element of M/I_1. Then $x/I_1 \oplus x/I_1 = x/I_1$ and so $(x \oplus x) \ominus x \leq a$ for some $a \in \mathcal{I}(M)$. From Proposition 3.2(v) and (iv) it follows that $x \oplus x \leq a \oplus x = x \oplus a$ and so, we have $x \oplus a \leq (x \oplus a) \oplus (x \oplus a) = (x \oplus x) \oplus a \leq (x \oplus a) \oplus a = x \oplus a$, that is $x \oplus a \in \mathcal{I}(M)$ which implies that $x \in I_1$. Hence, $x/I_1 = 0/I_1$ and consequently M/I_1 is strict. □

According to Theorem 5.8(3), every strict linearly ordered wPEMV-algebra is cancellative.

Question 5.10. Is every strict representable wPEMV-algebra cancellative? (For definition of a representable wPEMV-algebra see the beginning of Section 6.)

An ideal I of a wPEMV-algebra $(M; \vee, \wedge, \oplus, \ominus, \obar, 0)$, $I \neq M$, is said to be *maximal* if, for each ideal J of M, $I \subseteq J \subseteq M$ implies that $I = J$ or $J = M$. It is worthy of recalling that due to [21, Thm 4.16], every pseudo EMV-algebra admits at least one maximal ideal (not necessarily normal). This is not true for each wPEMV-algebra: Indeed, the wPEMV-algebra of the positive cone $\mathbb{R}^+$ has no maximal ideal.

Remark 5.11. If $M = (G^+; \vee, \wedge, \oplus, \ominus, \obar, 0)$ is a wPEMV-algebra of a positive cone, for the unital ℓ-group $(\mathbb{Z} \overrightarrow{\times} G, (1,0))$, we define a wPEMV-algebra $N = \Gamma_a(\mathbb{Z} \overrightarrow{\times} G, (1,0))$. Then the set $\{0\} \times G^+$ is a unique maximal and normal ideal of the wPEMV-algebra $(N; \vee, \wedge, \oplus, \ominus, \obar, (0,0))$ whose top element is $(1,0)$. Moreover, every element of N either belongs to M or is a difference $(1,0) \ominus x = x \obar (1,0)$ for some $x \in M$. We note that M is strict whereas N not.

Lemma 5.12. *Let I be a normal ideal of a wPEMV-algebra $(M; \vee, \wedge, \oplus, \ominus, \obot, 0)$. Then I is prime if and only if M/I is a linearly ordered wPEMV-algebra.*

Proof. Suppose that I is prime and let $x, y \in M$. By Proposition 3.2(xii), $x \ominus y \in I$ or $y \ominus x \in I$. From (5.3) it follows that $x/I \ominus y/I = (x \ominus y)/I = 0/I$ or $y/I \ominus x/I = (y \ominus x)/I = 0/I$ which means $x/I \leq y/I$ or $y/I \leq x/I$ (by Proposition 3.2(ix)). That is, M/I is a chain.

Conversely, let M/I be a linearly ordered wPEMV-algebra, and let $x \wedge y \in I$. Then $(x \wedge y)/I = 0/I$ and $x/I \leq y/I$ or $y/I \leq x/I$, so that $x/I = 0/I$ or $y/I = 0/I$. By (5.3), we have $x \in I$ or $y \in I$. □

We recall that if a wPEMV-algebra M possesses a top element, then $\lambda_1(x) := 1 \ominus x$ is said to be a *left complement* of x and the element $\rho_1(x) := x \obot 1$ is said to be a *right complement* of x. For simplicity, we write also $x^- = 1 \ominus x$ and $x^\sim = x \obot 1$. We remind that given $a \in M$, we set $\lambda_a(x) = a \ominus x$ and $\rho_a(x) = x \obot a$.

Lemma 5.13. (1) *Let M be a wPEMV-algebra without top element and let M be a subalgebra of a wPEMV-algebra N with top element. Take $x \in M$ and $a \in \mathcal{I}(M)$ such that $x \leq a$. Then $\lambda_a^2(x) = \lambda_1^2(x) \in M$ and $\rho_a^2(x) = \rho_1^2(x) \in M$.*

(2) *If M is a cancellative wPEMV-algebra without top element which is a subset of a cancellative wPEMV-algebra N with top element then, for each $x \in M$, we have $\lambda_1^2(x), \rho_1^2(x) \in M$.*

Proof. (1) First we note that since N is an associated wPEMV-algebra with top element, it is equivalent to a pseudo MV-algebra. Therefore, for each idempotent $b \in N$ and $z \in N$, we have $b \odot y = b \wedge y$, see [28, Prop 4.3].

Let $x \in M$ and $a \in \mathcal{I}(M)$ be an idempotent such that $x \leq a$. By Proposition 5.3(iii), we have $\lambda_1(x) = \lambda_a(x) \oplus \lambda_1(a)$, so that we have $\lambda_1^2(x) = \lambda_1(\lambda_a(x) \oplus \lambda_1(a)) = a \odot \lambda_1(\lambda_a(x)) = a \odot (\lambda_a^2(x) \oplus \lambda_1(a)) = a \wedge (\lambda_a^2(x) \vee \lambda_1(a)) = (a \wedge \lambda_a^2(x)) \vee (a \odot \lambda_1(a)) = a \wedge \lambda_a^2(x) = \lambda_a^2(x) \in M$. In the same way, we establish $\rho_1^2(x) = \rho_a^2(x) \in M$.

(2) According to Example 3.5, we can assume that M is a wPEMV-algebra of a positive cone G^+ of some ℓ-group G. Take $N = \Gamma_a(\mathbb{Z} \overrightarrow{\times} G, (1,0))$. Then for each $x = (0, g)$, where $g \in G^+$, we have $\lambda_1^2(x) = x = \rho_1^2(x) \in M$. □

Now, we present a similar representation result as the Basic Representation Theorem 2.2 for pseudo EMV-algebras.

Theorem 5.14. *Suppose a wPEMV-algebra M can be embedded into a wPEMV-algebra N_0 with top element. Then M either has a top element and so it is an associated wPEMV-algebra or it can be embedded into an associated wPEMV-algebra N with top element as a maximal and normal ideal of N. Moreover, every element* of N is either the image of $x \in M$ or is a right complement of the image of some element $x \in M$.

Proof. If M has a top element, the statement is trivial. Assume that the wPEMV-algebra M has no top element. We can assume that $M \subset N_0$ so that M is a proper wPEMV-subalgebra of N_0. We denote by $\ominus$ and $\oplus$ also the binary operations of N_0. Denote by $M^\sim = \{x \odot\!\!\sim 1 \mid x \in M\}$. We assert that $M \cap M^\sim = \emptyset$. Indeed, if $x \odot\!\!\sim 1 = y$ for some $x, y \in M$, then $1 = x \oplus (x \odot\!\!\sim 1) = x \oplus y$ which says $1 = x \oplus y \in M$, a contradiction.

Set $N = M \cup M^\sim$. We show that N is an associated wPEMV-subalgebra of N_0 which satisfies the conditions of our theorem.

Now, we define a binary operation $\odot$ on N_0 as $x \odot y := 1 \ominus ((y \odot\!\!\sim 1) \oplus (x \odot\!\!\sim 1)) = ((1 \ominus y) \oplus (1 \ominus x)) \odot\!\!\sim 1$, $x, y \in N_0$ (see (4.1)).

Claim. M is a proper ideal of N. Moreover, if $x, y \in M$, then $x \odot y \in M$, $x \ominus y = x \odot (1 \ominus y) \in M$, $x \odot\!\!\sim y = (x \odot\!\!\sim 1) \odot y \in M$, and $\lambda_1^2(x), \rho_1^2(x) \in M$.

First, we prove that M is an ideal of N. Since M is a wPEMV-algebra without top element, M is a proper subset of N. Clearly, M is closed under $\oplus$. Now, let $y \in N$ be such that $y \leq x$ for $x \in M$. Then y cannot be from $N \setminus M$, otherwise, $y = y_0 \odot\!\!\sim 1$ for some $y_0 \in M$, and $y_0 \odot\!\!\sim 1 \leq x$ yielding $1 \leq y_0 \oplus x$ which entails $y_0 \oplus x_0 = 1 \in M$, a contradiction. Whence $y \in M$.

Let $x, y \in M$. Since $x \odot y \leq x, y$, we conclude that $x \odot y \in M$. Using Theorem 4.3, we have $x \ominus y, x \odot\!\!\sim y \in M$. Now, we show that $\rho_1^2(x) \in M$ for each $x \in M$. Let $y = x^{\sim\sim}$ for some $y \in N$. Then $x^\sim = y^-$. If $y \in N \setminus N$, then $y = y_0 \odot\!\!\sim 1$, so that $x^\sim = y_0^{\sim -} = y_0 \in M$, a contradiction. Similarly, we have $\lambda_1^2(x) \in M$, which finishes the proof of the claim.

Clearly N contains M and 1. Let $x, y \in N$. We have three cases. Case (i): $x = x_0, y = y_0 \in M$. Then $x \vee y, x \wedge y, x \oplus y \in M$. Due to Theorem 4.3, we have $x \ominus y = x \ominus_1 y$ and $x \odot\!\!\sim y = x \odot\!\!\sim_1 y$ so that $x \ominus y, x \odot\!\!\sim y \in M \subset N$.

Case (ii): $x = x_0 \odot\!\!\sim 1$, $y = y_0 \odot\!\!\sim 1$ for some $x_0, y_0 \in M$. Then by (W7) and Theorem 3.2(vii), $x \vee y = (x_0 \odot\!\!\sim 1) \vee (y_0 \odot\!\!\sim 1) = (x_0 \wedge y_0) \odot\!\!\sim 1 \in N$, $x \wedge y = (x_0 \vee y_0) \odot\!\!\sim 1 \in N$ and $x \oplus y = (x_0 \odot\!\!\sim 1) \oplus (y_0 \odot\!\!\sim 1) = (y_0 \odot x_0) \odot\!\!\sim 1 \in N$ (using Claim). Due to Claim, we have $(x_0 \odot\!\!\sim 1) \odot\!\!\sim (y_0 \odot\!\!\sim 1) = \rho_1^2(x_0) \odot (y_0 \odot\!\!\sim 1) \leq \rho_1^2(x_0) \in M$. Analogously, $(x_0 \odot\!\!\sim 1) \ominus (y_0 \odot\!\!\sim 1) = (x_0 \odot\!\!\sim 1) \odot y_0 = x_0 \odot\!\!\sim y_0 \in M$.

Case (iii): We note that N_0 can be viewed also as a pseudo EMV-algebra with top element. In view of [21, Prop 3.4], we have for all $x, y \in N_0$, $x \ominus y = x \odot (1 \ominus y) = x \odot (1 \ominus (x \wedge y))$ and $x \odot\!\!\sim y = (x \odot\!\!\sim 1) \odot y = ((x \wedge y) \odot\!\!\sim 1) \odot y$.

Now, let $x = x_0$ and $y = y_0 \oslash 1$ for some $x_0, y_0 \in M$. Then

$$x \oplus y = x_0 \oplus (y_0 \oslash 1) = (y_0 \odot x_0^-) \oslash 1 = (y_0 \ominus x) \oslash 1 \in M^\sim,$$
$$y \oplus x = (y_0 \oslash 1) \oplus x_0 = ((1 \ominus x_0) \odot y_0) \oslash 1 \in M^\sim.$$

In addition, we have $x \wedge y = x_0 \wedge (y_0 \oslash 1) \leq x_0$, which yields $x \wedge y \in M$, and $x \vee y = x_0 \vee (y_0 \oslash 1) = (x_0^- \wedge y_0) \oslash 1 \in M^\sim$ since $x_0^- \wedge y_0 \leq y_0$ and hence $x_0^- \wedge y_0 \in M$. Moreover,

$$x \ominus y = x_0 \ominus y_0^\sim = x_0 \odot y_0 \in M,$$
$$y \ominus x = y_0^\sim \ominus x_0 = y_0^\sim \odot x_0^- = (\lambda_1^2(x_0) \oplus y_0) \oslash 1 \in M^\sim,$$
$$x \oslash y = x_0 \oslash y_0^\sim = x_0^\sim \odot y_0^\sim = (y_0 \odot x_0) \oslash 1 \in M^\sim,$$
$$y \oslash x = y_0^\sim \oslash x_0 = \rho_1^2(y_0) \odot x_0 \in M.$$

Now, we prove that M is a maximal and normal ideal of N. Since M is a wPEMV-algebra without top element, M is a proper subset of N. Now, let $y \in N \setminus M$, then $y = 1 \ominus y_0$ for some $y_0 \in M$. Then the ideal $Id(M, y)$ of N generated by M and $y_0 \oslash 1$ contains 1, so that $Id(M, y_0 \oslash 1) = N$ proving M is a maximal ideal of N.

Moreover, we prove that N is a normal ideal of N.

Let $y \oplus x_0 \in y \oplus M$, $y \in N_0$. It is sufficient to assume $y = y_0^\sim$ for some $y_0 \in M$. Then $y_0^\sim \oplus x_0 = (y_0^\sim \oplus x_0) \vee y_0^\sim = ((y_0^\sim \oplus x_0) \odot y_0) \oplus y_0^\sim$. Since $(y_0^\sim \oplus x_0) \odot y_0 \leq y_0 \in M$, we have $(y_0^\sim \oplus x_0) \odot y_0 \in M$, so that $y_0^\sim \oplus M \subseteq M \oplus y_0^\sim$. In a similar way, we prove the opposite inclusion. $\square$

Remark 5.15. The proof of Theorem 5.14 can be done also in an alternative way. Indeed, assume $M \subset N_0$ and define $N_1 = M \cup M^-$, where $M^- = \{\lambda_1(x) \mid x \in M\}$. From the proof of Claim of Theorem 5.14, we can show that if $x \in M$, then there is a unique $y \in M$ such that $x^\sim = y^-$ and vice versa. That is $M^\sim = M^-$. Then using the dual reasoning as in the proof of Theorem 5.14 we can show that N_1 is a wPEMV-algebra with top element and M is a maximal and normal ideal of N_1. In addition, the mapping $\phi : N \to N_1$ defined by $\phi(x) = \lambda_1^2(x)$, $x \in N$, is a wPEMV-isomorphism between N and N_1.

It is easy to see that if a wPEMV-algebra M without top element can be embedded into associated wPEMV-algebras N_1 and N_2 with top elements under wPEMV-algebra embeddings $\phi_i : M \to N_i$, $i = 1, 2$, such that $\phi_i(M)$ is a maximal and normal ideal of N_i and every element of N_i either belongs to $\phi_i(M)$ or it is a right (left) complement of some element from $\phi_i(M)$, then N_1 and N_2 are isomorphic associated wPEMV-algebras.

The associated wPEMV-algebra N with top element from Theorem 5.14 is said to be a wPEMV-algebra *representing* M. That is, if M is with top element, then $N = M$ and if M is topless, then N is with top element and M can be embedded into N as a maximal and normal ideal of N such that every element of N is either from the image of M or is a complement of some element from the image of M. We note that all wPEMV-algebras representing M are mutually isomorphic.

Since every wEMV-algebra, which is a commutative version of a wPEMV-algebra, is a subdirect product of linearly ordered wEMV-algebras, see [23, Prop 3.17, Thm 3.19], Theorem 5.14 is a generalization of an analogous representation theorem for wEMV-algebras from [23, Thm 3.20]. It would be interesting to show which classes of wPEMV-algebras M admit such a representation; that is when M can be embedded into an associated wPEMV-algebras N with top element as its maximal and normal ideal and every element of N either belongs to the image of M or is a complement of some element from the image of M.

In Section 6 of Part II, we show a positive answer for representable wPEMV-algebras. The general solution will be presented in Theorem 7.9 in Part II.

Acknowledgements

The authors are very indebted to anonymous referees for their careful reading, observations and suggestions which helped us to improve the presentation of the paper. In particular, we are indebted for a hint to look at wPEMV-algebras also as at integral GMV-algebras.

References

[1] P. Aglianò, G. Panti, *Geometrical methods in Wajsberg hoops*, J. Algebra **256** (2002), 352–374.

[2] P. Bahls, J. Cole, N. Galatos, P. Jipsen, C. Tsinakis, *Cancellative residuated lattices*, Algebra Universalis **50** (2003), 83–106.

[3] W. Blok, I.M.A. Ferreirim, *On the structure of hoops*, Algebra Universalis **43** (2000), 233–257.

[4] K. Blount, C. Tsinakis, *The structure of residuated lattices*, Inter. J. Algebra **13** (2003), 437–461.

[5] B. Bosbach, *Concerning semiclans*, Arch. Math. **37** (1981), 316–324.

[6] C.C. Chang, *Algebraic analysis of many-valued logics*, Trans. Amer. Math. Soc. **88** (1958), 467–490.

[7] C.C. Chang, *A new proof of the completeness of the Łukasiewicz axioms*, Trans. Amer. Math. Soc. **93** (1959), 74–80.

[8] R. Cignoli, I.M.L. D'Ottaviano and D. Mundici, *Algebraic Foundations of Many-Valued Reasoning*, Springer Science and Business Media, Dordrecht, 2000.

[9] P. Conrad, M.R. Darnel, *Generalized Boolean algebras in lattice-ordered groups*, Order **14** (1998), 295–319.

[10] M.R. Darnel, *Theory of Lattice-Ordered Groups*, Marcel Dekker, Inc., New York, Basel, Hong Kong, 1995.

[11] A. Di Nola, G. Georgescu, A. Iorgulescu, *Pseudo-BL-algebras: Part I*, Multi. Val. Logic **8** (2002), 673–714.

[12] A. Di Nola, G. Georgescu, A. Iorgulescu, *Pseudo-BL-algebras: Part II*, Multi. Val. Logic **8** (2002), 715–750.

[13] A. Dvurečenskij, *On pseudo MV-algebras*, Soft Computing **5** (2001), 347–354.

[14] A. Dvurečenskij, *Pseudo MV-algebras are intervals in ℓ-groups*, J. Austral. Math. Soc. **72** (2002), 427–445.

[15] A. Dvurečenskij, T. Vetterlein, *Pseudoeffect algebras. I. Basic properties*, Inter. J. Theor. Phys. **40** (2001), 685–701.

[16] A. Dvurečenskij, T. Vetterlein, *Pseudoeffect algebras. II. Group representation*, Inter. J. Theor. Phys. **40** (2001), 703–726.

[17] A. Dvurečenskij, T. Vetterlein, *Generalized pseudo-effect algebras*, In: Lectures on Soft Computing and Fuzzy Logic, Di Nola, A., Gerla, G., (Mds.), Physica-Verlag, Springer-Verlag Co., Berlin, 2001, pp. 89–111.

[18] A. Dvurečenskij, T. Vetterlein, *Algebras in the positive cone of po-groups*, Order **19** (2002), 127–146.

[19] A. Dvurečenskij, O. Zahiri, *On EMV-algebras*, Fuzzy Sets and Systems **373** (2019), 116–148.

[20] A. Dvurečenskij, O. Zahiri, *The Loomis–Sikorski theorem for EMV-algebras*, J. Austral. Math. Soc. **106** (2019), 200–234.

[21] A. Dvurečenskij, O. Zahiri, *Pseudo EMV-algebras. I. Basic Properties*, Journal of Applied Logics–IFCoLog Journal of Logics and their Applications **6** (2019), 1285–1327.

[22] A. Dvurečenskij, O. Zahiri, *Pseudo EMV-algebras. II. Representation and States*, Journal of Applied Logics–IFCoLog Journal of Logics and their Applications **6** (2019), 1329–1372.

[23] A. Dvurečenskij, O. Zahiri, *A variety containing EMV-algebras and Pierce sheaves of EMV-algebras*, Fuzzy Sets and Systems **418** (2021), 101–125. https://doi.org/10.1016/j.fss.2020.09.011

[24] A. Dvurečenskij, O. Zahiri, *Weak pseudo EMV-algebras. II. Representation and subvarieties*, **8** (2021), 2401–2433.

[25] D.J. Foulis, M.K. Bennett, *Effect algebras and unsharp quantum logics*, Found. Phys. **24** (1994), 1325–1346.

[26] L. Fuchs, *Partially Ordered Algebraic Systems*, Pergamon Press, Oxford-New York, 1963.

[27] N. Galatos, C. Tsinakis, *Generalized MV-algebras*, J. Algebra **283** (2005), 254–291.

[28] G. Georgescu and A. Iorgulescu, *Pseudo MV-algebras*, Multiple-Valued Logics **6** (2001), 193–215.

[29] G. Georgescu, L. Leuştean, V. Preoteasa, *Pseudo-hoops*, J. Multiple-Val. Logic Soft Comput. **11** (2005), 153–184.

[30] A.M.W. Glass, *Partially Ordered Groups*, World Scientific, Singapore, New Yersey, London, Hong Kong, 1999.

[31] P. Hájek, *Fuzzy logics with noncommutative conjunctions*. J. Logic Comput. **13** (2003), 469–479.

[32] D. Mundici, *Interpretation of AF C^*-algebras in Łukasiewicz sentential calculus*, J. Funct. Anal. **65** (1986), 15–63.

[33] J. Rachůnek, *A non-commutative generalization of MV-algebras*, Czechoslovak Math. J. **52** (2002), 255–273.

Received 25 May 2021

Weak Pseudo EMV-algebras. II. Representation and Subvarieties

Anatolij Dvurečenskij*
*Mathematical Institute, Slovak Academy of Sciences, Štefánikova 49, SK-814 73 Bratislava, Slovakia,
Palacký University Olomouc, Faculty of Sciences, tř. 17. listopadu 12, CZ-771 46 Olomouc, Czech Republic*
dvurecen@mat.savba.sk

Omid Zahiri
Tehran, Iran
zahiri@protonmail.com

Abstract

We define weak pseudo EMV-algebras which are a non-commutative generalization of weak EMV-algebras as well as of MV-algebras, pseudo MV-algebras, and of generalized Boolean algebras. In contrast to pseudo EMV-algebras, the class of wPEMV-algebras is a variety. We present basic properties and examples of wPEMV-algebras. The main aim is to show when a wPEMV-algebra can be embedded into a wPEMV-algebra N with top element, called a wPEMV-algebra representing M, as a maximal and normal ideal of N. The paper is divided into two parts. Part I studies wPEMV-algebras from the point of semiclans and generalized pseudo effect algebras. We describe congruences via normal ideals, and we show that a wPEMV-algebra possesses a representing one with top element.

Part II. It studies representable wPEMV-algebras and we show an equational base for them. Left and right unitizing automorphisms enable us to construct representing wPEMV-algebras. We present the Basic Representation Theorem. Finally, we study subvarieties of cancellative wPEMV-algebras, perfect wPEMV-algebras, weakly commutative wPEMV-algebras, and normal-valued wPEMV-algebras, respectively.

In this paper we continue in the study initiated in [22], where we introduced new algebraic structures called weak pseudo EMV-algebras and we have presented their basic properties. Sections, theorems, propositions, lemmas, examples, and equations are numbered in continuation of [22].

---

*Sponsored by the Slovak Research and Development Agency under contract APVV-16-0073 and the grant VEGA No. 2/0142/20 SAV

6 Representable wPEMV-algebras

We show that every representable wPEMV-algebras has a representing representable wPEMV-algebra with top element. Representable wPEMV-algebras form a variety and we show their equational base.

We say that a wPEMV-algebra M is *representable* if there is a system of linearly ordered wPEMV-algebras $(M_i)_{i \in I}$ such that there is an injective wPEMV-homomorphism $\varphi : M \to \prod_{i \in I} M_i$ with $\pi_i \circ \varphi(M) = M_i$ for all $i \in I$, where $\pi_i : \prod_{i \in I} M_i \to M_i$ is the i-th projection map. In other words, M is representable iff M is a subdirect product of linearly ordered wPEMV-algebras. Using Lemma 5.12 and standard techniques like in [19, Lem 5.9], we can show that a wPEMV-algebra M is representable if and only if there is a system $(P_i)_{i \in I}$ of normal prime ideals of M such that $\bigcap_{i \in I} P_i = \{0\}$.

Theorem 6.1. *Each representable wPEMV-algebra is a subalgebra of an associated representable wPEMV-algebra with top element.*

Proof. Let $(M; \vee, \wedge, \oplus, \ominus, \odot, 0)$ be a representable wPEMV-algebra. There is a system $(P_i)_{i \in I}$ of normal prime ideals of M such that $\bigcap_{i \in I} P_i = \{0\}$. Consider the embedding $f : M \to \prod_{i \in I} M/P_i$ defined by $f(x) = (x/P_i)_{i \in I}$. By Lemma 5.12, for each $i \in I$, M/P_i is a linearly ordered wPEMV-algebra and so by Theorem 5.8(1), M/P_i has a top element or it is strict. So, $I = I_1 \cup I_2$, where $I_1 = \{i \in I \mid M/P_i \text{ has a top element}\}$ and $I_2 = \{i \in I \mid M/P_i \text{ is strict}\}$. Clearly, $I_1 \cap I_2 = \emptyset$. Set $M_1 := \prod_{i \in I_1} M/P_i$ and $M_2 := \prod_{i \in I_2} M/P_i$. We know that $\prod_{i \in I} M/P_i$ and $M_1 \times M_2$ are isomorphic wPEMV-algebras, so we identify them.

(1) For each $i \in I_1$, M/P_i has a top element and it is an associated wPEMV-algebra so it can be viewed as a pseudo EMV-algebra. Thus, M_1 has a top element, too.

(2) For each $i \in I_2$, M/P_i is a linearly ordered strict wPEMV-algebra, hence by Theorem 5.8(2), it is a positive cone of an ℓ-group $(G; +, 0)$ which is linearly ordered. It follows from Remark 5.11 that M/P_i can be embedded into an associated wPEMV-algebra with top element. Thus, M_2 can be embedded into an associated representable wPEMV-algebra with top element, too.

From (1) and (2) we conclude that the representable wPEMV-algebra M is a subalgebra of an associated representable wPEMV-algebra with top element. □

It is useful to note that in the proof of Theorem 6.1, for each $i \in I_2$, $\mathcal{I}(M)$, the set of all idempotent elements of M, is a subset of P_i (otherwise, if $a \in \mathcal{I}(M) \setminus P_i$, then a/P_i is a non-zero idempotent element of M/P_i which is absurd). Hence $\mathcal{I}(M) \subseteq \bigcap_{i \in I_2} P_i$.

Now, we present a representation result for representable wPEMV-algebras.

Theorem 6.2. [Basic Representation Theorem for representable wPEMV-algebras] *Every representable wPEMV-algebra M either has a top element and so it is an associated wPEMV-algebra or it can be embedded into a representable wPEMV-algebra N with top element as a maximal and normal ideal of N. Moreover, every element of N is either the image of $x \in M$ or is a right complement of the image of some element $x \in M$.*

Proof. Due to Theorem 6.1, M can be embedded into a representable associated wPEMV-algebra N_0 with top element. Theorem 5.14 asserts that if we set $M^\sim = \{x \odot 1 \mid x \in M\}$, then $N = M \cup M^\sim$ is an associated wPEMV-algebra with top element that is a subalgebra of the representable associated wPEMV-algebra N_0. Let N_0 be a subdirect product of a system $\{N_i^0 \mid i \in I\}$ of linearly ordered wPEMV-algebras. Define $N_i := \pi_i(N)$, then N is a subdirect product of the system of linearly ordered wPEMV-algebras $\{N_i \mid i \in I\}$.

Consequently, applying Theorem 5.14, the result is established. □

The proof of the latter theorem is based on representability and it cannot be used for any general case of wPEMV-algebras. The general case of wPEMV-algebras will be solved in Theorem 7.9 below.

If $(M; \vee, \wedge, \oplus, \ominus, \odot, 0)$ is an associated wPEMV-algebra, then it corresponds to a pseudo EMV-algebra $(M; \vee, \wedge, \oplus, 0)$, see Example 3.3, and it can be embedded into a pseudo EMV-algebra with top element, see [20, Thm 6.4], so that for every associated wPEMV-algebra we have a representation theorem. Therefore, this suggests the following partial answer for representing wPEMV-algebras by Theorem 5.14.

Theorem 6.3. *Let a wPEMV-algebra M be a subdirect product of a system of associated wPEMV-algebras or of cancellative wPEMV-algebras. Then M has a PEMV-algebra with top element representing it.*

Proof. We will imitate the proofs of Theorems 6.1 and 6.2.

Thus, let $(M_i)_{i \in I}$ be a system of wPEMV-algebras such that every M_i is either an associated wPEMV-algebra or a cancellative wPEMV-algebra and let M be a subdirect product of $(M_i)_{i \in I}$. Let $I_1 = \{i \in I \mid M_i$ is an associated wPEMV-algebra$\}$ and $I_2 = \{i \in I \mid M_i$ is a cancellative wPEMV-algebra$\}$. Then $I_1 \cap I_2 = \emptyset$ and $I = I_1 \cup I_2$.

By [20, Thm 6.4], for each M_i with $i \in I_1$, there is an associated wPEMV-algebra N_i with top element in which M_i can be embedded as a maximal and normal ideal of N_i, and every element of N_i either belongs to the image of M_i or is a complement of some element from the image of M_i.

If $i \in I_2$, by Example 3.5, for each M_i, there is a unique (up to isomorphism) ℓ-group G_i such that M_i is isomorphic to the wPEMV-algebra of the positive cone G_i^+. Then G_i^+ can be embedded into $N_i = \Gamma_a(\mathbb{Z} \overrightarrow{\times} G_i, (1,0))$ and G_i^+ is a maximal and normal ideal of N_i and every element of N_i is either from G_i^+ or is a complement of some element from G_i^+.

Then $N_0 = \prod_{i \in I_1} N_i \times \prod_{i \in I_2} \Gamma_a(\mathbb{Z} \overrightarrow{\times} G_i, (1,0))$ is a wPEMV-algebra with top element and M can be embedded into N_0. Applying a general statement of Theorem 5.14, we get a desired result. $\square$

For every $z \in M$, we denote by a *polar* of z the set $z^\perp := \{x \in M \mid x \wedge y = 0\}$. There is a standard proof that $z^\perp$ is an ideal of M.

From the definition of a prime ideal we have that if P is a prime ideal, then the set $F = M \setminus P$ has the property $x \wedge y > 0$ for all $x, y \in F$, see also Lemma 5.4(ii). A subset F of $M \setminus \{0\}$ maximal under this condition is said to be an *ultrafilter*. There is a standard notion of a *minimal prime* ideal. Every prime ideal contains at least one minimal prime ideal. The proof of the following proposition follows the main steps of [26, Thm 2.20].

Proposition 6.4. *Let P be a proper ideal of a wPEMV-algebra M. The following statements are equivalent:*

(i) *P is a minimal prime ideal.*

(ii) *$M \setminus P$ is an ultrafilter.*

(iii)
$$P = \bigcup \{z^\perp : z \notin P\}. \tag{6.1}$$

(iv) *P is prime and for all $x \in P$, $x^\perp \not\subseteq P$.*

Proof. We prove only the implication (i) $\Rightarrow$ (ii). If $x, y \notin P$, then $x \wedge y \notin P$. Therefore, the Zorn Lemma guarantees that there is an ultrafilter F containing $M \setminus P$. We assert $F = M \setminus P$. Let $x \in F$, and $z \in x^\perp$. Then $z \wedge x = 0$ entails $z \notin F$, otherwise $z, x \in F$ and $z \wedge x > 0$, a contradiction. Hence,

$$x^\perp \subseteq M \setminus F \subseteq P.$$

Put $Q = \bigcup \{x^\perp : x \in F\}$. Then $Q \subseteq P$ and Q is an ideal of M. Indeed, clearly $0 \in Q$. Let $u, v \in Q$. There are $x, y \in F$ such that $u \in x^\perp$ and $v \in y^\perp$. Let a be an element of N such that $u \oplus v \oplus x \oplus y \leq a$. Due to [26, Prop 1.17] applied to the pseudo MV-algebra $([0,a]; \oplus_a, \lambda_a, \rho_a, 0, a)$, we have $x \wedge y \wedge (u \oplus v) =$

$x \wedge y \wedge (u \oplus_a v) \leq (x \wedge y \wedge u) \oplus_a (x \wedge y \wedge v) = 0$, that is, $u \oplus v \in (x \wedge y)^\perp$. Since $y \in F$ and $x \wedge y = (x \wedge y) \wedge y > 0$, we have $x \wedge y \in F$, so that $u \oplus v \in Q$. Finally, let $u \leq v \in Q$. Then there is $x \in F$ such that $v \in x^\perp$ which yields $u \in x^\perp$. Hence, Q is an ideal of M.

To prove Q is a prime ideal, let $u, v \in M$ be such that $u \wedge v = 0$ and let $u \notin Q$. Since F is an ultrafilter, $u \in F$ and $v \in u^\perp \subseteq Q$. In other words, we have established Q is prime such that it is contained in P and minimality of P entails $Q = P$.

The proof of the other implications is identical to the proof of ones in [26, Thm 2.20]. $\square$

Lemma 6.5. *A wPEMV-algebra M is representable if and only if every polar $z^\perp$ is a normal ideal.*

Proof. Assume that M is representable and let it be a subdirect product of a system $(M_i)_{i \in I}$ of linearly ordered wPEMV-algebras. Without loss of generality we can assume that M is a subalgebra of $\prod_{i \in I} M_i$. Let $x, y, z \in M$ and $x = (x_i)_i$, $y = (y_i)_i$ and $z = (z_i)_i$. Suppose that $x \ominus y \in z^\perp$. Then $(x_i \ominus y_i) \wedge z_i = 0$ for each i. If $z_i = 0$, then $(y_i \ominus x_i) \wedge z_i = 0$. If $z_i > 0$, then $0 = x_i \ominus y_i = (x_i \ominus y_i) \wedge z_i$ so that Proposition 3.2(ix) yields $x_i \leq y_i$ and whence, $(y_i \ominus x_i) \wedge z_i = 0$. The converse implication is analogous and this establishes that $z^\perp$ is normal.

Conversely, let each $z^\perp$ be a normal ideal of M. Using (6.1), we see that every minimal prime ideal P is normal, and the intersection of all minimal prime ideals is the zero ideal $\{0\}$. Hence, M is a subdirect product of linearly ordered wPEMV-algebras $(M/P)_P$, where P is a minimal prime ideal of M, and whence, M is representable. $\square$

Inspired by [28, Thm 3.4], we present equations characterizing representable wPEMV's. We note that the binary operation $\oplus_a$ can be expressed as $x \oplus_a y = (x \oplus y) \wedge a$, so it is expressible using the language of wPEMV-algebras.

Theorem 6.6. *A pseudo EMV-algebra M is representable if and only if, for all $x, y, z, a \in M$ such that $x, y, z \leq a$, we have*

$$\left(x \ominus y\right) \wedge \left((z \oplus_a (y \ominus x)) \ominus z\right) = 0,$$
$$\left(y \ominus x\right) \wedge \left(z \ominus ((x \ominus y) \oplus_a z)\right) = 0.$$

As a corollary, the class of representable wPEMV-algebras is a subvariety of wPEMV-algebras.

Proof. If M is a linear wPEMV-algebra, then due to Theorem 5.8, it either has a top element or is cancellative, and in either case it satisfies both equations. Consequently, every representable pseudo EMV-algebra satisfies the equations.

Conversely, assume that M satisfies the above equations. We show that every polar $x^\perp$, $x \in M \setminus \{0\}$, is normal. Let $y \in x^\perp$ and assume $x \oplus y \oplus z \leq a$ for some $a \in M$. If $[0, a]$ is assumed as a pseudo MV-algebra, we have $a \ominus x = \lambda_a(x) = \lambda_a(x) \oplus_a 0 = \lambda_a(x) \oplus_a (x \wedge y) = (\lambda_a(x) \oplus_a x) \wedge (\lambda_a(x) \oplus_a y) = a \wedge (\lambda_a(x) \oplus_a y) = \lambda_a(x) \oplus_a y$. In a similar way, $a \ominus y = \lambda_a(y) = \lambda_a(y) \oplus_a x$, whence by Theorem 4.3, $x = \rho_a(y) \odot_a x = y \oslash x$ and $y = \rho_a(x) \odot_a y = x \oslash y$. Similarly, $y = y \odot_a \lambda_a(x) = y \ominus x$ and $x = x \odot_a \lambda_a(y) = x \ominus y$. Then we have

$$x \wedge \Big((z \oplus_a y) \ominus z\Big) = 0,$$
$$x \wedge \Big(z \oslash (y \oplus_a z)\Big) = 0,$$

which implies $(z \oplus_a y) \ominus z \in x^\perp$ and $z \oslash (y \oplus_a z) \in x^\perp$. Since $y \oplus z = y \oplus_a z = (y \oplus_a z) \vee z = z \oplus_a (z \oslash (y \oplus_a z))$ and $z \oplus y = z \oplus_a y = ((z \oplus_a y) \ominus z) \oplus z$, we see that $x^\perp$ is a normal ideal. By Lemma 6.5, M is representable.

If in the above two formulas we change x by $x \wedge a$, y by $y \wedge a$ and z by $z \wedge a$, then we have two identities for all $x, y, z, a \in M$, so that the class of representable wPEMV-algebras forms a subvariety. $\square$

In the language of integral GMV-algebras, from [25, Thm 2.2], [2], we have that the class of representable integral GMV-algebras is a variety; compare with [22, Prop 4.9].

We denote by Repr the class of representable wPEMV-algebras.

As a corollary, we have the following characterization of representable wPEMV-algebras.

Proposition 6.7. *A wPEMV-algebra M is representable if and only if, for each $a \in M$, the associated wPEMV-algebra $M_a = ([0, a]; \vee, \wedge, \oplus_a, \ominus_a, \oslash, 0)$ is representable.*

Proof. According to Theorem 4.3, $([0, a]; \oplus_a, \lambda_a, \rho_a, 0, a)$ is a pseudo MV-algebra and M_a is an associated wPEMV-algebra with top element a. Applying equations from Theorem 6.6, we see that M_a is representable for every $a \in M$. Equivalently, M is representable. $\square$

By Theorem 6.6, we know that the class of representable wPEMV-algebras is a variety containing the class of associated representing wPEMV-algebras. Moreover, by Theorem 6.2, each representable wPEMV-algebra is a subalgebra of a representable wPEMV-algebra with top element. Therefore, we have the following corollaries.

Corollary 6.8. *The variety* Repr *of representable wPEMV-algebras is the least variety containing the class of representable associated wPEMV-algebras.*

Corollary 6.9. *Every subdirectly irreducible representable wPEMV-algebra is linearly ordered.*

Proof. Due to Proposition 5.6, we know that there is a one-to-one correspondence between normal ideals and congruences. Therefore, the variety of wPEMV-algebras is congruence-distributive, since the set of normal ideals of M is a distributive lattice. To prove the latter statement, let A, B, C be three normal ideals of M. Let $A \vee B$ denote the normal ideal generated by $A \cup B$, similarly we have $A \wedge C = A \cap B$. Then $(A \vee B) \wedge C = (A \vee B) \cap C \supseteq (A \wedge C) \vee (B \vee C)$. Let $x \in C$ belong to the ideal generated by $A \cup B$, then $x \leq a \oplus b$, where $a \in A$, $b \in B$. We have $x \leq a \oplus b = (a \wedge (c \ominus b)) + b$, where $c = a \oplus b$. Since the GPEA $(M; +, 0)$ satisfies RDP$_2$, $x = a_1 + b_1$ for some $a_1 \leq a$ and $b_1 \leq b$ (RDP$_0$). Then $a_1 \in A \cap C$ and $b_1 \in B \cap C$ which implies x belongs to the ideal generated by $(A \cap C) \cup (B \cap C) \subseteq (A \cap C) \vee (B \cap C)$. Hence, $(A \vee B) \cap C \subseteq (A \cap C) \vee (B \cap C)$.

Using RDP$_2$, we can show also the second distributive law $(A \wedge B) \vee C = (A \vee C) \wedge (B \vee C)$: Clearly, $(A \wedge B) \vee C \subseteq (A \vee C) \wedge (B \vee C)$. Now, let x belong to the ideals generated by $A \cup C$ and $B \cup C$, respectively. Then $x \leq a \oplus c_1$ and $x \leq b \oplus c_2$, where $a \in A$, $b \in B$ and $c_1, c_2 \in C$. As in the latter paragraph, we can assume that $x = a + c_1 = b + c_2$. Using RDP$_2$, there are four elements $c_{11}, c_{12}, c_{21}, c_{22}$ such that $a = c_{11} + c_{12}$, $c_1 = c_{21} + c_{22}$, $b = c_{11} + c_{21}$ and $c_2 = c_{21} + c_{22}$. Then $c_{11} \in A \cap B$, $c_{12} \in A \cap C$, $c_{21} \in B \cap C$ and $c_{22} \in C$. We have $x = c_{11} + c_{12} + c_{21} + c_{22}$ and therefore, x belongs to the ideal generated by $(A \cap B) \cup (A \cap C) \cup (B \cap C) \cup C = (A \cap B) \cup C$, and then $x \in (A \wedge B) \vee C$.

Applying Jónsson's Lemma [6, Cor 6.9] on congruence-distributive varieties, we conclude that every subdirectly irreducible representable wPEMV-algebra has to be linearly ordered. □

The latter result has a counterpart in integral GMV-algebras just before [25, Thm 2.2].

Proposition 6.10. *The variety of wPEMV-algebras is arithmetical.*

Proof. Let us consider the following ternary terms: $M(x, y, z) = (x \vee y) \wedge (x \vee z) \wedge (y \vee z)$ and $m(x, y, z) = ((x \ominus y) \oplus z) \wedge ((z \ominus y) \oplus x) \wedge (x \vee z)$. Then $M(x, y, z)$ is a majority Mal'cev term proving in another way as it was done in the latter corollary that wPEMV is a congruence-distributive variety. For $m(x, y, z)$, we have

(1) $m(x,y,x) = ((x \ominus y) \oplus x) \wedge ((x \ominus y) \oplus x) \wedge (x \vee x) = x.$

(2) $m(x,y,y) = ((x \ominus y) \oplus y) \wedge ((y \ominus y) \oplus x) \wedge (x \vee y) = (x \vee y) \wedge x \wedge (x \vee y) = x.$

(3) $m(y,y,x) = ((y \ominus y) \oplus x) \wedge ((x \ominus y) \oplus y) \wedge (y \vee x) = x \wedge (x \vee y) \wedge (x \vee y) = x.$

This proves the variety of wPEMV-algebras is arithmetical. □

7 wPEMV-algebras and Generalized Pseudo Effect Algebras

We will study a close relation between wPEMV-algebras and generalized pseudo effect algebras. We introduce left and right unitizing automorphisms which enable us to construct a representing wPEMV-algebra for a wPEMV-algebra without top element.

It is well known, see e.g. [17, Thm 8.8, Prop 8.7], that if a PEA E is a lattice PEA satisfying RDP$_2$, we can define a total binary operation $\oplus$ by $x \oplus y := (x \wedge y^-) + y$ for all $x, y \in E$ such that $(E; \oplus, ^-, ^\sim, 0, 1)$ is a pseudo MV-algebra. In particular, if E is a lattice GPEA with RDP$_2$, then every $([0,a]; \oplus_a, \lambda_a, \rho_a, 0, a)$ $(a \in E)$, where

$$x \oplus_a y := (x \wedge \lambda_a(y)) + y = (x \wedge (a \setminus y)) + y, \quad x, y \in [0, a], \qquad (7.1)$$

is a pseudo MV-algebra.

Let $(M; +, 0)$ be a GPEA. A one-to-one mapping $\phi : M \to M$ is said to an *automorphism* if $x + y$ exists iff $\varphi(x) + \varphi(y)$ exists, and $\varphi(x + y) = \varphi(x) + \varphi(y)$, $\varphi^{-1}(x+y) = \varphi^{-1}(x) + \varphi^{-1}(y)$. Inspired by [23, Def 4.1], we introduce the following notions. An automorphism $\varphi_\lambda : M \to M$ is said to be *left unitizing* iff, for all $x, y \in M$, $\varphi_\lambda(x) + y$ exists iff $y + x$ is defined. Dually, an automorphism $\varphi_\rho : M \to M$ is said to be *right unitizing* iff $x + \varphi_\rho(y)$ is defined iff $y + x$ is defined. If M is a wPEMV-algebra, we say that it admits a right and left unitizing automorphisms if it does the GPEA $(M; +, 0)$.

If M is a PEA, that is, M possesses a top element, then due to [23, Lem 2.5, Lem 2.7], M admits a unique left unitizing automorphism and a unique right unitizing automorphism, namely, $\varphi_\lambda(a) = a^{--}$ and $\varphi_\rho(a) = a^{\sim\sim}$, $a \in M$.

Lemma 7.1. *If M is an associated wPEMV-algebra, then M admits both left and right unitizing automorphisms and they are inverses of each other. The same is true if M is a cancellative wPEMV-algebra; in this case the left and right unitizing automorphisms coincide.*

Proof. (1) In an associated wPEMV-algebra M, each element $x \in M$ is dominated by some idempotent element $a \in \mathcal{I}(M)$ and every associated wPEMV-algebra is a subalgebra of some associated wPEMV-algebra with top element (see e.g. [20, Thm 6.4]). According to Lemma 5.13, if $x \le a, b \in \mathcal{I}(M)$, then $\rho_a^2(x) = \rho_b^2(x)$ and $\lambda_a^2(x) = \lambda_b^2(x)$, where $\lambda_a(x) = a \ominus x$ and $\rho_a(x) = x \oslash a$, so the mappings $\varphi_\rho(x) := \rho_a^2(x)$ and $\varphi_\lambda(x) := \lambda_a^2(x)$, $x \in M$, do not depend on the choice of $a \in \mathcal{I}(M)$.

Now, we show that φ_ρ and φ_λ are automorphisms. Let $x, y \in M$ be such that $x+y$ is defined in M. There is an idempotent $a \in M$ such that $x + y \le a$, and therefore, $x, y \in [0, a]$, where $([0, a]; +_a, 0, a) = ([0, a]; +, 0, a)$ is a PEA with negations $x^{-a} = \lambda_a(x)$ and $x^{\sim a} = \rho_a(x)$. Since if $x + y$ is defined, then $x = (x \oplus y) \ominus y \le a \ominus y$ giving $x +_a y$ is defined in $[0, a]$ and whence $x + y = x +_a y$. Therefore, $x + y + (x+y)^{\sim a} = a$ and $y + (x+y)^{\sim a} = x^{\sim a}$. Thus $y + (x+y)^{\sim a} + x^{\sim a \sim a} = a$, which yields $y^{\sim a} = (x+y)^{\sim a} + x^{\sim a \sim a}$. Therefore, $a = y^{\sim a} + y^{\sim a \sim a} = (x+y)^{\sim a} + x^{\sim a \sim a} + y^{\sim a \sim a}$, which implies $(x+y)^{\sim a \sim a} = x^{\sim a \sim a} + y^{\sim a \sim a}$, that is, $\varphi_\rho(x+y) = \varphi_\rho(x) + \varphi_\rho(y)$. Clearly, $\varphi_\rho(0) = 0$ and $\varphi_\rho(a) = a$. The same is true for φ_λ. Since $\varphi_\lambda(\varphi_\rho(x)) = x$, φ_ρ is an automorphism.

We note that $\varphi_\lambda(x) + y$ is defined iff $y \le \varphi_\lambda^{\sim a}(x) = x^{-a}$ iff $y + x$ is defined in $[0, a]$. Then φ_λ is a left unitizing automorphism. Analogously, the same is true for φ_ρ.

(2) Now, let M be a cancellative wPEMV-algebra. There is an ℓ-group G such that $M \cong G^+$ and M is a maximal and normal ideal of $\Gamma_a(\mathbb{Z} \overrightarrow{\times} G, (1, 0))$. Due to Lemma 5.13(2), we have $\lambda_1^2(x) = \rho_1^2(x)$ for each $x \in M$ so that $\varphi_\lambda(x) = \lambda_1^2(x) = \rho_1^2(x) = \varphi_\rho(x)$, and similarly as in (1), we can show that φ_λ is an automorphism that is both left and right unitizing. □

In the next proposition, we show that the existence of a right (left) unitization automorphism on a wPEMV-algebra M without top element is a necessary condition for M to have a representing wPEMV-algebra with top element.

Proposition 7.2. *Let M be a wPEMV-algebra without top element. If it has a representing wPEMV-algebra N with top element, then M admits left and right unitization automorphisms.*

Proof. If M is subalgebra of a representing wPEMV-algebra N with top element 1, then $\varphi_\lambda(x) = \lambda_1^2(x)$, and $\varphi_\rho(x) = \rho_1^2(x)$, $x \in M$, are unitization automorphisms on M, see Lemma 7.1. □

Theorem 7.3. *Let $(M; \vee, \wedge, \oplus, \ominus, \oslash, 0)$ be a wPEMV-algebra without top element. Let M admit a left unitizing automorphism φ_λ and let $(M; +, 0)$ be the GPEA derived from M according to Theorem 4.6. There is a PEA $(N; \hat{+}, 0, 1)$ with RDP_2 such that*

GPEA $(M; +, 0)$ can be embedded into $(N; \hat{+}, 0, 1)$ as a maximal and normal PEA-ideal of N.

Proof. According to Lemma 4.4, on the set M there is a partial operation $+$ such that $x+y$ exists iff there is $a \in M$ such that if $x \oplus y \leq a$ and $x \leq \lambda_a(y)$. Then $x+y := x \oplus y$ and $(M; +, 0)$ is a GPEA without top element satisfying RDP$_2$. Let M' be a set of the same cardinality as M such that $M \cap M' = \emptyset$. Let $\eta : M \to M'$ be a bijection. Then $M' = \eta(M)$. If we set $N := M \cup \eta(M)$, we define a partial operation $\hat{+}$ on N that is an extension of the original partial operation $+$ on M as follows:

(i) If $x, y \in M$, then $x+y$ is defined in M iff $x\hat{+}y$ is defined in N, in which case $x\hat{+}y = x + y$.

(ii) If $x = x_0$ and $y = \eta(y_0)$, where $x_0, y_0 \in M$, then $x \hat{+} y$ is defined in N iff $x_0 \leq y_0$, in which case $x \hat{+} y := \eta(y_0 \setminus x_0)$.

(iii) If $x = \eta(x_0)$ and $y = y_0$, where $x_0, y_0 \in M$, then $x \hat{+} y$ is defined in N iff $\varphi_\lambda(y_0) \leq x_0$, in which case $x \hat{+} y := \eta(\varphi_\lambda(y_0) / x_0)$, where φ_λ is a left unitizing automorphism of the PGEA $(M; +, 0)$.

(iv) If $x = \eta(x_0)$, $y = \eta(y_0)$, where $x_0, y_0 \in M$, then $x\hat{+}y$ is not defined.

Since the mapping φ_λ is an automorphism of M into itself such that $\varphi_\lambda(u)+v$ is defined in M iff $v+u$ is defined in M, according to [23, Thm 4.2], we conclude that $(N; \hat{+}, 0, 1)$, where $1 := \eta(0)$, is a PEA. Moreover, $\varphi_\lambda(x) = x^{--}$ and $\varphi_\rho(x) = x^{\sim\sim}$, $x \in M$.

In addition, for $x_0, y_0, z_0 \in M$, we have

$$x_0 \hat{+} y_0^\sim = z_0^\sim \Leftrightarrow z_0 + x_0 = y_0 \text{ and } x_0^\sim \hat{+} y_0 = z_0^\sim \Leftrightarrow y_0^{--} + z_0 = x_0. \quad (7.2)$$

Claim 1. The PEA $(N; \hat{+}, 0, 1)$ is a lattice.

To prove the claim, we exhibit 10 cases.

(1) $\eta(a) = a^\sim$, $\varphi_\lambda(a) = a^{--}$, $\eta(a)^- = a$, $\eta(a)^\sim = a^{\sim\sim}$ and $\eta(\varphi_\lambda(a)) = \varphi_\lambda^\sim(a) = a^-$ for each $a \in M$, and $0^- = 1$.

(2) We have also $a^{\sim\sim} = \varphi_\rho(a) \in M$ for each $a \in M$. Indeed, there is $b \in M$ such that $\varphi_\rho(a) = b$. Then $a = \varphi_\lambda(b) = b^{--}$ which yields $a^{\sim\sim} = b = \varphi_\rho(a)$.

(3) If $x \leq c \in M$, then $1 = c \hat{+} c^\sim = x \hat{+} \rho_c(x) \hat{+} c^\sim$ which entails $x^\sim = \rho_c(x) \hat{+} c^\sim$. In a similar way we can show that $x^- = c^- \hat{+} \lambda_c(x)$.

(4) If $c \in \mathcal{I}(M)$, then $c^{--} = c = c^{\sim\sim}$ and $\varphi_\rho(c) = c^{\sim\sim} = c$ which gives $c^\sim = c^-$.

The set M is a maximal and normal PEA-ideal in N, see [23, Thm 3.3(vii)]. Then $x \leq y$ in M for $x, y \in M$ iff $x \leq_N y$ in N. Therefore, we can denote the partial order $\leq_N$ on N also simply as $\leq$ that is, as it is used in M.

(5) If $v, x, y \in M$, then $v \leq x^\sim$ implies $x \hat{+} v$ is defined in N and by definition of $\hat{+}$, we have $x \hat{+} v = x + v \in M$, so that $v \leq \rho_c(x)$ for each $c \geq x + y$, see Lemma 4.4. In the same way, we have $v \leq y^-$ implies $v \leq \lambda_c(y)$ for each $c \geq v + y$.

(6) If $c \in \mathcal{I}(M)$, then $c \wedge_N c^\sim = 0 = c \wedge_N c^-$ and $c \vee_N c^\sim = 1 = c \vee_N c^-$. Indeed, let $x \leq c, c^\sim$. Then $x \in M$ and by (5), $c + x$ is defined in M and due to Proposition 3.2(xiii), $c + x = c \vee x = c$ which implies $x \leq \rho_c(c) = 0$. Similarly $c \wedge_N c^- = 0$.

Now, let $c^\sim, c \leq y \in N$. Then $y = y_0^\sim$ for a unique $y_0 \in M$. Since then $y_0 \leq c, c^\sim = c^-$, we have $y_0 = 0$ and $y = 1$.

(7) Now we show that if $a, b \in M$, then $a \wedge b = a \wedge_N b$ and $a \vee b = a \vee_N b$. Let $z_0 \in M$ be such that $z_0^\sim \leq a, b$. But this is impossible because then $z_0^\sim \in M$ while M is an ideal, so that $1 = z_0 \hat{+} z_0^\sim \in M$ which is a contradiction. Whence, $a \wedge b = a \wedge_N b$.

Now let $a, b \leq z_0^\sim$, where $z_0 \in M$. Then $z_0 + a$ exists in N and since $z_0, a \in M$, $z_0 + a$ exists in M, similarly $z_0 + b$ is defined in M. By Lemma 4.4, there are elements $c_1, c_2 \in M$ such that $a + z_0 \leq c_1$, $a \leq \rho_{c_1}(z_0)$ and $b + z_0 \leq c_2$, $b \leq \rho_{c_2}(z_0)$. If we take $c \geq c_1, c_2$, $c \in M$, then $a, b, z_0 \leq c$ and $a, b \leq \rho_c(z_0)$. So that $a \vee b \leq \rho_c(z_0)$ and $z_0 + (a \vee b)$ is defined in M. (W10) entails $(z_0 + a) \vee (z_0 + b) = (z_0 \oplus a) \vee (z_0 \oplus b) = z_0 \oplus (a \vee b) = z_0 + (a \vee b)$. Hence, $(a \vee b) \leq z_0^\sim$ which shows $a \vee b = a \vee_N b$.

In addition, it is straightforward that if $a, b \in M$, then $a^\sim \vee_N b^\sim = (a \wedge b)^\sim$. To show $a^\sim \wedge_N b^\sim = (a \vee b)^\sim$, let $x \in M$ be such that $x^\sim \leq a^\sim, b^\sim$. Then clearly $x^\sim \leq (a \vee b)^\sim$. If $x \in M$ and $x \leq a^\sim, b^\sim$, then $a + x$ and $b + x$ are defined in M and by Lemma 4.4, there is $c \in M$ with $c \geq a + x, b + x$ such that $a, b \leq \lambda_c(x)$. Then $a \vee b \leq \lambda_c(x)$ and $x \leq \rho_c(a \vee b) \leq (a \vee b)^\sim$. Both cases give $a^\sim \wedge_N b^\sim = (a \vee b)^\sim$.

(8) We show that $x^\sim \wedge_N y$ exists in N and it belongs to M, moreover, $x^\sim \wedge_N y = x \ominus (x \oplus y)$. According to (W4), $x \ominus (x \oplus y) \leq y$ and on the other side, $x \ominus (x \oplus y) = x / (x \oplus y) \leq x / 1 = x^\sim$. Now, let $a \leq x^\sim, y$, then $a \in M$ and $x + a$ is defined in M. Therefore, $x \ominus (x \oplus y) \geq x \ominus (x \oplus a) = a$ proving $x \ominus (x \oplus y) = x^\sim \wedge_N y \in M$.

Moreover, if $c \geq x \oplus y$, we assert that $x^\sim \wedge_N y = \rho_c(x) \wedge y$. Indeed, put $c_0 = x \oplus y$ and let $c \geq c_0$. Then $x^\sim \wedge y = x \ominus (x \oplus y) = \rho_{c_0}(x) \wedge y \leq \rho_c(x) \wedge y \leq (x \ominus 1) \wedge_N y = x^\sim \wedge y$.

Since $x^- = x_0^\sim$ for some $x_0 \in M$, we have that $x^- \wedge_N y$ exists in N and it belongs to M. Similarly as in the last paragraph, we can show that $x^- \wedge y = (y \oplus x) \ominus x$, and for each $c \geq x \oplus y$, we have $x^- \wedge y = \lambda_c(x) \wedge y \in M$.

(9) The partial order $\leq_N$ is a lattice order on N. From (7) and (8), we conclude that N is a $\wedge_N$-lattice. Now, if $x, y \in M$, we show $x^\sim \vee_N y$ exists and it belongs to $N \setminus M$. Let $a \geq x^\sim, y$. Then $a^- \leq x, y^-$ and by (8), $a^- \leq x \wedge y^- = (x \oplus y) \ominus y$ and

$a \geq ((x \oplus y) \ominus y)^\sim = x^\sim \vee_N y$.

In other words, due to (7) and (8), we see that N is also a $\vee_N$-lattice.

(10) If $a, b \in M$, $a \leq b$, then $a^\sim \setminus b^\sim = a \odot b = a \,/\, b$, $b^\sim \,/\, a^\sim = (b \ominus a)^{\sim\sim}$ and $b^- \setminus a^- = b \setminus a$. Indeed, for the first equality, let $d \in M$ be such that $a^\sim \setminus b^\sim = d$ and $a^\sim = d + b^\sim$ so that by (7.2), we have $a + d = b$ and $d = a \setminus b = a \odot b$. In a similar way, we prove the second equality.

Claim 2. *The PEA* $(N; \hat{+}, 0, 1)$ *satisfies* RDP_2.

To prove RDP_2, we exhibit four cases. To make calculations more easier, since $\hat{+}$ is an extension of the partial addition $+$ on M, we will write $\hat{+} = +$.

(I) Let $a + b = c + d$ for $a, b, c, d \in M$. This case is evident.

(II) Let $a^\sim + f = b^\sim + g$, where $a, b, f, g \in M$. Then there is $z \in M$ such that $a^\sim + f = z^\sim = b^\sim + g$. From (7.2), we have $\varphi_\lambda(f) + z = a$ and $\varphi_\lambda(g) + z = b$, so that $\varphi_\lambda(f) \,/\, a = z = \varphi_\lambda(g) \,/\, b$. Since M is directed, there is $d \in M$ such that $a, b \leq d$. Then $d \setminus (\varphi_\lambda(f) \,/\, a)$ is defined in M. We assert $d \setminus (\varphi_\lambda(f) \,/\, a) = (d \setminus a) + \varphi_\lambda(f)$. Indeed, we have $d = (d \setminus a) + (a \setminus \varphi_\lambda(f)) + \varphi_\lambda(f)$, which entails that the element $x = (d \setminus a) + \varphi_\lambda(f)$ is defined in M. Moreover,

$$d \setminus a = x \setminus \varphi_\lambda(f)$$
$$d = (x \setminus \varphi_\lambda(f)) + a$$
$$= (x \setminus \varphi_\lambda(f)) + \varphi_\lambda(f) + (\varphi_\lambda(f) \,/\, a)$$
$$= x + (\varphi_\lambda(f) \,/\, a)$$
$$x = d \setminus (\varphi_\lambda(f) \,/\, a).$$

In the same way we can prove that $d \setminus (\varphi_\lambda(g) \,/\, b) = (d \setminus b) + \varphi_\lambda(g)$. Since $d \setminus (\varphi_\lambda(f) \,/\, a) = d \setminus (\varphi_\lambda(g) \,/\, b)$, we have

$$(d \setminus a) + \varphi_\lambda(f) = (d \setminus b) + \varphi_\lambda(g).$$

Using RDP_2 holding in M, there are four elements $c_{11}, c_{12}, c_{21}, c_{22} \in M$ such that the following RDP_2 table holds

	c_{11}	c_{12}
$d \setminus a$		
$\varphi_\lambda(f)$	c_{21}	c_{22}
	$d \setminus b$	$\varphi_\lambda(g)$

.

From this table we have

$$d \setminus a = c_{11} + c_{12}, \quad d \setminus b = c_{11} + c_{21}$$
$$d = c_{11} + c_{12} + a, \quad d = c_{11} + c_{21} + b$$
$$c_{12} + a = c_{21} + b,$$

which implies

	$a^\sim$	$(c_{12}+a)^\sim$	$\varphi_\rho(c_{12})$
	f	$\varphi_\rho(c_{21})$	$\varphi_\rho(c_{22})$
		$b^\sim$	g

Clearly, $\varphi_\rho(c_{12}) \wedge \varphi_\rho(c_{21}) = 0$.

(III) $f + a^\sim = z^\sim = g + b^\sim$ for $a, b, f, g, z \in M$.

From (7.2), we have $z + f = a$ and $z + g = b$ which implies $a \setminus f = z = b \setminus g$. There is an element $d \in M$ such that $a, b \leq d$. Using ideas from the proof of part (II), we have

$$(a \setminus f) / d = f + (a / d) = (b \setminus g) / d = g + (b / d)$$

which gives an RDP_2 table holding in M

	f	d_{11}	d_{12}
	a/d	d_{21}	d_{22}
		g	b/d

Since $a + d_{21} = b + d_{12}$, we have an RDP_2 table for case (II)

	f	d_{11}	d_{12}
	$a^\sim$	d_{21}	$(a+d_{21})^\sim$
		g	$b^\sim$

which is an RDP_2 table.

(IV) Let $a^\sim + f = z^\sim = g + b^\sim$ for $a, b, g, h, z \in M$. We find an RDP_2 decomposition.

Using (7.2), we have $\varphi_\lambda(f) + z = a$ and $z + g = b$, so that $\varphi_\lambda(f) / a = z = b \setminus g$. Let c be an element of M such that $g \oplus a, f \oplus b, z, \varphi_\lambda(f) \leq c$. By the representation theorem, see [14], there is a unital ℓ-group (G_c, c) such that $([0, c]; \oplus_c, \lambda_c, \rho_c, 0, c) \cong \Gamma(G_c, c)$. Therefore, in the group G_c, we have

$$-\varphi_\lambda(f) + a = b - g$$
$$a + g = c + f - c + b$$
$$-c + a + g = f - c + b. \quad (*)$$

Since $\rho_c(a) \leq a^\sim$ and $\rho_c(b) \leq b^\sim$, we are looking for an RDP table in the form

	$a^\sim$	c_{11}	$\widetilde{c_{12}}$
	f	c_{21}	c_{22}
		g	$b^\sim$

where $c_{11}, c_{12}, c_{21}, c_{22} \in M$. Using (8), if we put $c_{11} := a^\sim \wedge g$, then $c_{11} = \rho_c(a) \wedge g$, $c_{22} := b^\sim \wedge f = \rho_c(b) \wedge f$ and $c_{21} = c_{11} / g = (\rho_c(a) \wedge g) / g$. Then $c_{11} = \rho_c(a) \odot_c (a \oplus_c g)$ and $c_{22} = \rho_c(b) \odot_c (b \oplus_c f)$. We show that $c_{21} = f \setminus c_{22} = f \setminus (\rho_c(b) \wedge f)$. Using (4.3), we have $c_{21} = \rho_c^2(a) \odot_c g = (\rho_c^2(a) - c + g) \vee 0 = (-c + a + g) \vee 0$.

On the other side, put $c'_{21} := f \setminus c_{22} = f \odot_c b = (f - c + b) \vee 0$, so that by $(*)$, $c_{21} = c'_{21}$.

For c_{12} we have $a^\sim = c_{11} + c_{12}^\sim$, that is $a + c_{11} = c_{12}$ and $c_{12} = a + (\rho_c(a) \wedge g) = a \oplus_c g$. Similarly, if $c'_{12} \in M$ is such that $(c'_{12})^\sim + c_{22} = b^\sim$, then $c_{22}^{\overline{\sim}} + b = c'_{12}$, which gives

$$\begin{aligned}
c'_{12} &= \varphi_\lambda(c_{22}) + b = \varphi_\lambda(\rho_c(b) \wedge f) + b \\
&= (\lambda_c(b) \wedge \varphi_\lambda(f)) + b = (\varphi_\lambda(f) \wedge \lambda_c(b)) + b \\
&= (\varphi_\lambda(f) + b) \wedge (\lambda_c(b) + b) \\
&= (\varphi_\lambda(f) + b) \wedge c = (\varphi_\lambda(f) \oplus b) \wedge c \\
&= \varphi_\lambda(f) \oplus_c b.
\end{aligned}$$

From $\varphi_\lambda(f) / a = b \setminus g$ we obtain in the unital ℓ-group (G_c, c)

$$\begin{aligned}
-\varphi_\lambda(f) + a &= b - g \\
a + g &= \varphi_\lambda(f) + b \\
a \oplus_c g &= \varphi_\lambda(f) \oplus_c b,
\end{aligned}$$

which proves $c_{12} = c'_{12}$, so that RDP holds in N.

Using (8), we have $c_{12}, c_{21} \in [0, c]$ so that using Proposition 3.2(ii), we have $c_{21} \wedge_N c_{12}^\sim = (\rho_c^2(a) \odot_c g) \wedge (a \oplus_c g)^\sim = (\rho_c^2(a) \odot_c g) \wedge (a \oplus_c g)^\sim \wedge c = (\rho_c^2(a) \odot_c g) \wedge (\rho_c(a \oplus_c g)) = (\rho_c(\rho_c(a)) \odot_c g) \wedge (\rho_c(g) \odot_c \rho_c(a)) = 0$.

Hence, the table gives also an RDP$_2$ decomposition. □

We note that Theorem 7.3 holds also if a wPEMV-algebra M admits a right unitizing automorphism.

The latter theorem entails that every wPEMV-algebra with an unitizing automorphism can be represented by an associated wPEMV-algebra with top element.

Now, we show that the existence of a left (right) unitizing automorphism is a necessary and sufficient condition in order that a wPEMV-algebra admits a representing wPEMV-algebra with top element.

Theorem 7.4. *Every wPEMV-algebra M without top element and admitting a left unitizing automorphism can be embedded into an associated wPEMV-algebra N with* *top element as a maximal and normal ideal of N. Moreover, every element of N is either the image of $x \in M$ or is a right complement of the image of some element $x \in M$.*

Proof. According to Theorem 7.3, there is a PEA N with RDP_2 such that the GPEA $(M;+,0)$ derived from M can be embedded into a GPEA $(N;+_N,0,1)$ as a maximal and normal PEA-ideal of N. Without loss of generality, we can assume $M \subset N$. Since N satisfies RDP_2, then N can be converted into a pseudo EMV-algebra $(N;\oplus_N,^-,^\sim,0,1)$, see [17, Thm 8.8], [16], where the pseudo MV-operation $\oplus_N$ is defined by $x \oplus_N y = (x \wedge_N y^-) +_N y$, $x,y \in N$. It is clear that M is a wPEMV-ideal of N. In particular, if $x, y \in M$, let $c \in M$ be such that $x \oplus y \leq c$. Then by (8) of the proof of Theorem 7.3, $x \oplus_N y = (x \wedge_N y^-) +_N y = (x \wedge_N \lambda_c(y)) +_N y = (x \wedge \lambda_c(y)) +_c y = (x +_c y) \wedge (\lambda_c(x) +_c y) = (x +_c y) \wedge c = x \oplus y$, that is the binary operation $\oplus_N$ is an extension of $\oplus$.

Moreover, if $x,y \in N$, then $x \odot_N y = (y^- \oplus_N x^-)^\sim = (y^\sim \oplus_N x^\sim)^-$, $x \ominus_N y = x \odot_N y^- = x \setminus (x \wedge_N y)$ and $x \oslash_N y = x^\sim \odot_N y = (x \wedge_N y) / y$. Therefore, if $x,y \in M$, then $x \ominus y = x \ominus_N y$ and $x \oslash y = x \oslash_N y$.

Altogether, we have proved that the wPEMV-algebra $(M; \vee, \wedge, \oplus, \ominus, \oslash, 0)$ can be embedded into the associated wPEMV-algebra $(N; \vee_N, \wedge_N, \oplus_N, \ominus_N, \oslash_N)$ with top element. In addition, the wPEMV-algebra N satisfies the conditions of Theorem. That is, M is a maximal and normal ideal of N, and every element of N is either the image of some element of M or is a right complement of the image of some element of M. □

Now, we emphasize the importance of the both left and right unitizing automorphisms for existence of a representing wPEMV-algebra with top element.

Theorem 7.5. *Let M be a wPEMV-algebra without top element. Then there is a wPEMV-algebra N representing M with top element if and only if M possesses a left (right) unitizing automorphism.*

Proof. If M possesses a right unitizing automorphism, Theorem 7.3 shows that M has a representing wPEMV-algebra N with top element. Conversely, let N be a wPEMV-algebra with top element representing M. For simplicity, let $M \subset N$. For every $x \in M$, we set $\varphi_\lambda : x \mapsto x^{--}$ and $\varphi_\rho(x) = x^{\sim\sim}$. Then $x^- \in N \setminus M$. If $x^{--} \in N \setminus M$, there is $y \in M$ such that $x^{--} = y^-$ which entails $x^- = y \in M$, a contradiction, therefore, $x^{--} \in M$. Applying [23, Lem 2.7, Lem 2.5], we have that φ_λ and φ_ρ are automorphisms of M that are right unitizing and left unitizing automorphisms. □

It remains an open question whether every wPEMV-algebra without top element does possess a right (left) unitization automorphism. A positive answer to this question will be posed at the end of this section.

In the language of integral GMV-algebras, see [22, Prop 4.9], the latter question can be formulated in an equivalent way whether every integral GMV-algebra can be embedded into a bounded integral GMV-algebra, compare with [22, Thm 5.14].

Now, we present a corollary of the latter result for a special class of wPEMV-algebras. We say that a wPEMV-algebra M is *weakly commutative* if, given $x, y \in M$, $x+y$ is defined in M iff so is defined $y+x$. For example, every wEMV-algebra (i.e. $\oplus$ is commutative) and every cancellative wPEMV-algebra is weakly commutative. We note that the second case shows that the weakly commutativity does not imply that $\oplus$ is commutative.

Corollary 7.6. *Every weakly commutative wPEMV-algebra M without top element admits an associated wPEMV-algebra N with top element representing M. Moreover, for each $x \in N$, $x^\sim = x^-$ and N is also weakly commutative.*

Proof. If M is weakly commutative, then the identity mapping φ is both a left and right unitization automorphism. Existence of N follows from Theorem 7.5.

Now, let $x \in M$. Since the identity mapping is both a left and right unitization automorphism, by [23, Lem 2.5, Lem 2.7] it is a unique unitization automorphism. Then $x^{\sim\sim} = x = x^{--}$ which gives $x^\sim = x^-$. On the other hand, $(x^\sim)^- = x = (x^\sim)^\sim$, so that $y^- = y^\sim$ holds for each $y \in N$. Now, let, for $x, y \in N$, $x + y$ be defined in N. Then $x \leq y^- = y^\sim$ giving $y + x$ is defined in N, and vice versa. □

We finish this section with a result saying how we can embed a wPEMV-algebra without top element into a bounded distributive lattice.

We present a simple but useful lemma.

Lemma 7.7. *Let M be a wPEMV-algebra. Let $x, y \in M$ and let $c \geq c_0 := x \oplus y$. Then $x \ominus (x \oplus y) = \rho_{c_0}(x) \wedge y = \rho_c(x) \wedge y$. Similarly, if $c \geq c'_0 := y \oplus x$, then $(y \oplus x) \ominus x = \lambda_{c'_0}(x) \wedge y = \lambda_c(x) \wedge y$.*

Proof. Due to (W4), we have $x \oplus (x \ominus c) = c$. Moreover, $x \ominus (x \oplus y) \leq x \ominus c$ and by (W3), $x \ominus (x \oplus y) \leq y$, so that $x \ominus (x \oplus y) \leq \rho_c(x) \wedge y$. Now, let $a \leq \rho_c(x), y$. Then $x + a = x +_c a = x \oplus_c a = x \oplus a$ is defined in the pseudo MV-algebra $([0, c]; \oplus_c, \lambda_c, \rho_c, 0, c)$, so that $x \ominus (x \oplus y) \geq x \ominus (x \oplus a) = a$ establishing $x \ominus (x \oplus y) = \rho_c(x) \wedge y$.

In the dual way, we show the second equality. □

Assume that M is a wPEMV-algebra without top element. Define the set $N = M \cup M'$, $M \cap M' = \emptyset$, where M' is a set of the same cardinality as M. Let $\eta : M \to M'$ be a bijection, i.e. $M' = \eta(M)$. We will write also $\eta(x) = x^\sim$, $x \in M$. On the set N, we define a relation $\leq_N$ in the following way: Let $x, y \in M$. Then

(i) $x \leq_N y$ iff $x \leq y$,
(ii) $x^\sim \leq_N y^\sim$ iff $y \leq x$,
(iii) $x \leq_N y^\sim$ iff $x = y \obar (y \oplus x)$,
(iv) $y^\sim \not\leq_N x$.

Theorem 7.8. *Let M be a wPEMV-algebra without top element. Then the relation $\leq_N$ on N is a lattice ordering that is an extension of $\leq$ on M and N under this order is a bounded distributive lattice.*

Proof. (1) The relation $\leq_N$ is an extension of $\leq$ and it is reflexive and antisymmetric. We show that it is also transitive. It is sufficient to exhibit the following two cases (a) $x \leq y$, $y \leq_N z^\sim$ and (b) $x \leq_N y^\sim$, $y^\sim \leq_N z^\sim$. Case (a): We have $x \leq y = z \obar (z \oplus y)$. Put $c_y = z \oplus y$ and $c_x = z \oplus x$. Then $c_x \leq c_y$ and by Lemma 7.7, we get $\rho_{c_y}(z) \wedge x = \rho_{c_x}(z) \wedge x = (z \obar (z \oplus x)) \wedge x = z \obar (z \oplus x)$. On the other hand, $\rho_{c_y}(z) \wedge x = (z \obar (z \oplus y)) \wedge x = y \wedge x$, that is $x \leq_N z^\sim$.

Case (b): In this case, we have $x = y \obar (y \oplus x)$ and $z \leq y$. Define $c_y = y \oplus x$ and $c_z = z \oplus x$, so that $c_z \leq c_y$. Therefore,

$$x = y \obar (y \oplus x) \leq z \obar (y \oplus x)$$
$$x = (z \obar (y \oplus x)) \wedge x$$
$$= \rho_{c_y}(z) \wedge x = \rho_{c_z}(z) \wedge x \quad \text{(by Lemma 7.7)}$$
$$= (z \obar (z \oplus x)) \wedge x = z \obar (z \oplus x),$$

i.e. $x \leq_N z^\sim$ and $\leq_N$ is a partial order on N such that $0 \leq_N x, y^\sim \leq_N 0^\sim =: 1_N$, $x, y \in M$, and

$$x \leq c \in M \implies \rho_c(x) \leq_N x^\sim. \tag{7.3}$$

(2) Now, we show that N is a $\wedge_N$-lattice. Take $x, y \in N$. Then clearly (i) $x \wedge_N y = x \wedge y$. (ii) $x^\sim \wedge_N y^\sim = (x \vee y)^\sim$. We have $(x \vee y)^\sim \leq x^\sim, y^\sim$. Choose $z \in M$ and let $z^\sim \leq_N x^\sim, y^\sim$ and $z \geq x \vee y$, so that $z^\sim \leq (x \vee y)^\sim$. Now, let $z \leq_N x^\sim, y^\sim$. Then $x \obar (x \oplus z) = z = y \obar (y \oplus z)$ and define $c_x = x \oplus z$, $c_y = y \oplus z$, and $c = (x \vee y) \oplus z$. This yields $c = c_x \vee c_y$ and applying Lemma 7.7, we get $(x \vee y) \obar ((x \vee y) \oplus z) = \rho_c(x \vee y) \wedge z = \rho_c(x) \wedge \rho_c(y) \wedge z = \rho_{c_x}(x) \wedge \rho_{c_y} \wedge z = z$. Both subcases entail $x^\sim \wedge_N y^\sim = (x \vee y)^\sim$.

Finally (iii): We show $x \wedge_N y^\sim = y \obar (y \oplus x)$. By (W3) $y \obar (y \oplus x) \leq_N x$. In addition, if $c = y \oplus x$, then $y \obar (y \oplus x) = \rho_c(y) \leq_N y^\sim$, see (7.3). Now, take

$z \leq_N x, y^\sim$. Then $z = y \ominus (y \oplus z) = \rho_{c_0}(y) \leq \rho_{c_0}(y)$, where $c_0 = y \oplus z$, which guarantees $y +_{c_0} z = y \oplus_{c_0} z = y \oplus z$ exists in $[0, c_0]$. Consequently, $y \ominus (y \oplus x) \geq y \ominus (y \oplus z) = z$ which implies $y \ominus (y \oplus x) = x \wedge_N y^\sim$. Applying Lemma 7.7, we have also $x \wedge_N y^\sim = x \wedge \rho_c(y)$, where $c \geq y \oplus x$.

(3) In what follows, we show that $\leq_N$ is a lattice order on N. Since N is a $\wedge_N$-lattice, we show that it is also a $\vee_N$-lattice. (i) Clearly $x^\sim \vee_N y^\sim = (x \wedge y)^\sim$. (ii) We have $x \vee_N y = x \vee y$. If $z \geq x, y$, then $z \geq x \vee y$, that is $z \geq_N x \vee y$. Now, let $z^\sim \geq_N x, y$. Then $x = z \ominus (z \oplus x)$, $y = z \ominus (z \oplus y)$. If we put $c_x = z \oplus x$ and $c_u = z \oplus y$, then $c := z \oplus (x \vee y) = c_x \vee c_y$. Check

$$z \ominus (z \oplus (x \vee y)) = z \ominus ((z \oplus x) \vee (z \oplus y))$$
$$= \rho_c(z) \wedge (x \vee y) = (\rho_c(z) \wedge x) \vee (\rho_c(z) \wedge y)$$
$$= x \vee y,$$

where we have used Lemma 7.7.

In (iii) we assert that $x^\sim \vee_N y = ((x \oplus y) \ominus y)^\sim$. By (W3), we have $(x \oplus y) \ominus y \leq x$, so that $x^\sim \leq_N ((x \oplus y) \ominus y)^\sim$. On the other hand, $(x \oplus y) \ominus y = \lambda_c(y)$, where $c = x \oplus y$. Then $\rho_c((x \oplus y) \ominus y) = y$. Property (7.3) entails $y = \rho_c((x \oplus y) \ominus y) \leq_N ((x \oplus y) \ominus y)^\sim$. That is $x^\sim, y \leq ((x \oplus y) \ominus y)^\sim$.

Now, let $z \geq_N x^\sim, y$. Then $z \in M'$, so that we assume $z^\sim \geq_N x^\sim, y$ which gives $z \leq x$ and $y = z \ominus (z \oplus y)$. Set $c = z \oplus y$. Then $y = \rho_c(z)$ and $z = \lambda_c(y) \leq \lambda_c(y)$. Therefore, $z +_c y = z \oplus_c y = z \oplus y$ is defined in $[0, c]$. This entails $(x \oplus y) \ominus y \geq (z \oplus y) \ominus y = z$ and $((x \oplus u) \ominus y)^\sim \leq_N z^\sim$ which finishes the proof of (iii). Using Lemma 7.7, we have also $x^\sim \vee_N y = (\lambda_c(y) \wedge x)^\sim$, where $c \geq x \oplus y$.

Altogether $\leq_N$ is a lattice order on N.

(4) We show that $(a \vee_N b) \wedge_N c = (a \wedge_N c) \vee_N (b \wedge_N c)$. Assume $x, y, z \in M$. It is clear that this distributivity law holds for $a = x$, $b = y$, $c = z$ and for $a = x^\sim$, $b = y^\sim$, $c = z^\sim$. We verify the following cases.

Case (i): $a = x$, $b = y$ and $c = z^\sim$. Set $c = z \oplus (x \vee y) = c_x \vee c_y$ with $c_x = z \oplus x$ and $c_y = z \oplus y$. By (2)(iii), we have

$$(x \vee_N y) \wedge_N z^\sim = (x \vee y) \wedge_N z^\sim = \rho_c(z) \wedge (x \vee y)$$
$$= (\rho_c(z) \wedge x) \wedge (\rho_c(z) \wedge y)$$
$$= (\rho_{c_x}(z) \wedge x) \vee (\rho_{c_y}(z) \wedge y)$$
$$= (x \wedge_N z^\sim) \vee_N (y \wedge_N z^\sim).$$

Case (ii): $a = x$, $b = y^\sim$, $c = z$. Due to Lemma 7.7, (2)(iii), and (3)(iii), take a sufficiently large $c \in M$. Then

$$(x \vee_N y^\sim) \wedge_N z = (\lambda_c(x) \wedge y)^\sim \wedge_N z$$
$$= \rho_c(\lambda_c(x) \wedge y) \wedge z = (x \wedge z) \vee (z \wedge \rho_c(y))$$
$$= (x \wedge_N z) \vee_N (y^\sim \wedge_N z).$$

Case (iii): $a = x^\sim$, $b = y^\sim$, $c = z$. Take again sufficiently large $c \in M$. Then

$$(x^\sim \vee_N y^\sim) \wedge_N z = (x \wedge y)^\sim \wedge_N z = \rho_c(x \wedge y) \wedge z$$
$$= (\rho_c(x) \vee \rho_c(y)) \wedge z = (\rho_c(x) \wedge z) \vee (\rho_c(y) \wedge z)$$
$$= (x^\sim \wedge_N z) \vee_N (y^\sim \wedge_N z).$$

Case (iv): $a = x^\sim$, $b = y$, $c = z^\sim$. Take a sufficiently large c, then

$$(x^\sim \vee_N y) \wedge_N z^\sim = (\lambda_c(y) \wedge x)^\sim \wedge_N z^\sim = ((\lambda_c(y) \wedge x) \vee z)^\sim$$
$$= ((\lambda_c(y) \vee z) \wedge (x \vee z))^\sim = (\lambda_c(y) \vee z)^\sim \vee_N (x \vee z)^\sim$$
$$= (y \wedge_N z^\sim) \vee_N (x^\sim \wedge_N z^\sim).$$

(5) In the same way, we can show that also the second distributivity law $(a \wedge_N b) \vee_N c = (a \vee_N c) \wedge_N (b \vee_N c)$ holds for all $a, b, c \in N$. $\square$

In what follows, we introduce two new algebras: bricks and clans. According to [4], an algebra $(B; \ominus, \oslash, 1)$ of type $(2,2,0)$ is said to be a *brick* if the following holds:

(B1) $(a \oslash a) \oslash b = b = b \ominus (a \ominus a)$;

(B2) $a \oslash (b \ominus c) = (a \oslash b) \ominus c$;

(B3) $a \ominus (b \oslash a) = (b \ominus a) \oslash b$;

(B4) $1 \ominus (a \oslash 1) = a$.

Due to [4], $a \ominus a = b \ominus b = a \oslash a = b \oslash b$ for all $a, b \in B$. The element $a \ominus a$ is denoted as 0. If we define $a \leq b$ iff $a \ominus b = 0$, then $0 \leq x \leq 1$ for each $x \in B$ and B is a $\wedge$-semilattice, $a \wedge b = a \ominus (b \oslash a) = (b \ominus a) \oslash b$, and due to [31, Lem 3.1], it is also a lattice. For example, every pseudo MV-algebra can be viewed also as a brick.

An algebra $(C, \ominus, \oslash)$ of type $(2,2)$ is called a *cone algebra* (cone, for simplicity), if

(C1) $(a \oslash a) \oslash b = b = b \ominus (a \ominus a)$;

2419

(C2) $(a \obar b) \ominus c = a \obar (b \ominus c)$;

(C3) $a \ominus (b \obar a) = (b \ominus a) \obar b$;

(C4) $(a \obar b) \obar (a \obar c) = (b \obar a) \obar (b \obar c)$;

(C5) $(c \ominus a) \ominus (b \ominus a) = (c \ominus b) \ominus (a \ominus b)$.

According to [4, p. 65], every brick is a cone algebra. Similarly as for bricks, we put $0 = a \ominus a = a \obar a$ which is the same for each $a \in C$. The most important example of cone algebras is the positive cone of any ℓ-group G with $g \ominus h = (g-h) \vee 0$ and $g \obar h = (-g+h) \vee 0$, $g,h \in G^+$.

Now, we are able to present a main result, representation theorem for wPEMV-algebras which generalizes Theorem 6.2 for representable wPEMV-algebras.

Theorem 7.9 (Basic Representation Theorem for wPEMV-algebras). *Every wPEMV-algebra M either has a top element and so it is an associated wPEMV-algebra or it can be embedded into an associated wPEMV-algebra N with top element as a maximal and normal ideal of N. Moreover, every element of N is either the image of $x \in M$ or is a right complement of the image of some element $x \in M$.*

Proof. Let $(M; \vee, \wedge, \oplus, \ominus, \obar, 0)$ be a wPEMV-algebra. If M is with top element, the statement is trivial. Thus, let M have no top element. It is easy to verify that $(M; \ominus, \obar)$ is a clan algebra. Due to the Second Embedding Theorem of Bosbach, [4], every cone algebra can be embedded into some brick B. By [31, Thm 3.3] as well as by [29, Cor] and [30, Cor], bricks and pseudo MV-algebras are term equivalent.

We note that in wPEMV-algebras, using (W3) and (v) of Proposition 3.2, we have: Given $x, y \in M$, $x \oplus y$ can be defined from $\ominus$ and $\obar$, respectively, in this way: $x \oplus y = \max\{z \in M \mid z \ominus y \leq x\} = \max\{z \in M \mid x \obar z \leq y\}$. The same is true in bricks thought as pseudo MV-algebras or as bounded associated wPEMV-algebras. In other words, the wPEMV-algebra M can be embedded into an associated wPEMV-algebra B with top element.

Applying the general result of Theorem 5.14 from the first part, we conclude the assertion in question. □

As a corollary of the latter theorem, we have that every wPEMV-algebra admits a unique left unitization automorphism and a unique right unitization automorphism, see Theorem 7.5.

Finally, we present a generalization of Corollary 6.8:

Corollary 7.10. *The variety of wPEMV-algebras is the least variety containing the class of associated wPEMV-algebras.*

Proof. It follows the analogous steps as Corollary 6.8. □

8 Subvarieties of wPEMV-algebras

In the section, we define different kinds of subvarieties of wPEMV-algebras, namely, cancellative, perfect, commutative, weakly commutative, and normal-valued ones and for some of them we present an equational base.

The variety wPEMV of wPEMV-algebras has the following subvarieties: (i) O, the variety of wPEMV-algebras satisfying the equation $x = 0$; it contains only the trivial algebra, (ii) Bool, the variety of wPEMV-algebras satisfying the equation $x \oplus x = x$. (iii) wEMV, the wPEMV-algebras satisfying the equation $x \oplus y = y \oplus x$. Then O $\subsetneq$ Bool $\subsetneq$ wEMV $\subsetneq$ wPEMV. We note that according to [21, Thm 3.23], the variety of commutative wPEMV-algebra wEMV has only countably many subvarieties. In this section, we show that the variety wPEMV has uncountably many subvarieties.

We introduce two important unital ℓ-groups, for more details, see [8, Chap 6]. We denote by $\mathrm{Aut}(\mathbb{R})$ the set of order automorphisms on $\mathbb{R}$, i.e. the set of all continuous strictly increasing mappings from $\mathbb{R}$ onto $\mathbb{R}$. Let $u \in \mathrm{Aut}(\mathbb{R})$ be the translation $u(t) = t + 1$, $t \in \mathbb{R}$, and

$$\mathrm{BAut}(\mathbb{R}) := \{g \in \mathrm{Aut}(\mathbb{R}) \colon \exists\, n \geq 1, u^{-n} \leq g \leq u^n\}.$$

Then $\mathrm{Aut}(\mathbb{R})$ generates the variety of ℓ-groups, [8, Thm 38.23], $(\mathrm{BAut}(\mathbb{R}), u)$ is a unital ℓ-group and $\Gamma(\mathrm{BAut}(\mathbb{R}), u)$ generates the variety of pseudo MV-algebras, see [15, Cor 4.9]. By the way, if some identity containing only $\vee, \wedge, \ominus, \oslash$ holds in the positive cone wPEMV-algebra $\mathrm{Aut}(\mathbb{R})^+$, then it holds in every wPEMV-algebra, see the note just after Theorem 4.6.

Theorem 8.1. *Let Can be the class of cancellative wPEMV-algebras. Then Can is a proper subvariety of the variety wPEMV-algebras and a wPEMV-algebra M belongs to Can if and only if M satisfies the identities*

$$x = (x \oplus y) \ominus y, \quad x = y \oslash (y \oplus x).$$

If Can_c is the class of commutative cancellative wPEMV-algebras, then it is an atom in the lattice of subvarieties of commutative wPEMV-algebras, and it has a single generator, $\mathbb{Z}^+$. Moreover, the subvariety Can contains uncountably many subvarieties of cancellative wPEMV-algebras and the wPEMV-algebra $\mathrm{Aut}(\mathbb{R})^+$ generates the variety Can. The subvariety Bool is an atom in the lattice of subvarieties of the variety of wPEMV-algebras.

Proof. Let a wPEMV-algebra M satisfy the two identities. Assume $x \oplus y = x \oplus z$. Then $y = x \oslash (x \oplus y) = x \oslash (x \oplus z) = z$. Similarly, $x_1 \oplus y = x_2 \oplus y$ implies $x_1 = x_2$, i.e. M is cancellative. Conversely, let M be cancellative. According to Theorem 5.8(2),

M is isomorphic to a wPEMV-algebra of the positive cone G^+ of some ℓ-group G. Hence, M satisfies both equations.

Let $\mathcal{LG}$ be the variety of all ℓ-groups, and let $\Psi : \mathcal{LG} \to \mathsf{Can}$ be a mapping defined by $\Psi(G) = (G^+; \vee, \wedge, \oplus, \ominus, \odot, 0)$ assigning for every ℓ-group G its wPEMV-algebra of the positive cone G^+. Using the Nakada theorem, [24, Prop X.1], it is possible to show that Ψ defines a categorical equivalence between the category of cancellative wPEMV-algebras and the category of ℓ-groups. If $\mathcal{V}$ is a subvariety of ℓ-groups, then $\Psi(\mathcal{V})$ is a subvariety of cancellative wPEMV-algebras. Let $\mathcal{V}_1$ and $\mathcal{V}_2$ be two different subvarieties of ℓ-groups and take an ℓ-group $G \in \mathcal{V}_1 \setminus \mathcal{V}_2$. Then $\Psi(G) \in \Psi(\mathcal{V}_1) \setminus \Psi(\mathcal{V}_2)$. Therefore, the mapping Ψ is injective.

Let V be a subvariety of cancellative wPEMV-algebras. Then V is defined by a set of equations Σ in the language of $\vee, \wedge, \oplus, \ominus, \odot$. Since $x \oplus y = x + y$, $x \ominus y = (x - y) \vee 0$, $x \odot y = (-x + y) \vee 0$, equations in Σ use the language of ℓ-groups with variables in G^+. For each $(\sigma = \tau) \in \Sigma$, let $(\sigma' = \tau')$ be the equation in the language of ℓ-groups obtained from $(\sigma = \tau)$ by replacing $x \oplus y$ by $x + y$, $x \ominus y$ by $(x - y) \vee 0$ and $x \odot y$ by $(-x + y) \vee 0$. Then $G^+ \models (\sigma = \tau)$ iff for all $g_1, \ldots, g_n \in G^+$, $\sigma(g_1, \ldots, g_n) = \tau(g_1, \ldots, g_n)$. This is true iff $G^+ \models (\sigma' = \tau')$, equivalently, for all $g_1, \ldots, g_n \in G^+$, we have $\sigma'(g_1, \ldots, g_n) = \tau'(g_1, \ldots, g_n)$. Now, let $(\sigma'' = \tau'')$ be the equation obtained by replacing each variable x in $\sigma' = \tau'$ be $x \vee 0$. Whence, $G^+ \models (\sigma' = \tau')$ iff $G \models (\sigma'' = \tau'')$. Let $\mathcal{V}$ be the subvariety of ℓ-groups defined by all equations $(\sigma'' = \tau'')$ where $(\sigma = \tau) \in \Sigma$. Therefore, $\Psi(\mathcal{V}) = \mathsf{V}$ and Ψ is surjective.

It is well-known that the variety of ℓ-groups has uncountably many different subvarieties, see [27, Thm 10.K], [8], consequently, the variety Can contains uncountably many mutually different subvarieties of cancellative wPEMV-algebras.

The fact that Can_c is an atom in the variety of wEMV-algebras was established in [21, Thm 3.22] and the notes just before the present theorem establish that $\mathrm{Aut}(\mathbb{R})^+$ generates Can.

Finally, let V be any non-zero subvariety such that $\mathsf{O} \not\subseteq \mathsf{V} \subseteq \mathsf{Bool}$. Since Bool contains only those commutative wPEMV-algebras M such that $M = \mathcal{I}(M)$, then every M has a representing wEMV-algebra where each element is idempotent. Then V contains only commutative wPEMV-algebras K such that $K = \mathcal{I}(K)$. Consequently, every $K \in \mathsf{V}$ is an associated wPEMV-algebra, so it is a q-subvariety of EMV-algebras in the sense of [18], and in view of the proof of [18, Thm 5.22], there is a one-to-one relationship between q-subvarieties of EMV-algebras and subvarieties of MV-algebras. Therefore, $\mathsf{Bool} = \mathsf{V}$. □

In the language of integral GMV-algebras, the above equational base for cancellative integral GMV-algebras was established also in [1].

We note that the variety of Boolean algebras is the least non-trivial subvariety of

MV-algebras, because every proper subvariety of MV-algebras is determined only by finitely many equations using a single variable, see [11]. Therefore, it is contained in each non-zero subvariety of MV-algebras. We have to note that this is not true for the subvariety Bool and any non-zero subvariety of wEMV-algebras, because each non-zero generalized Boolean algebra is not cancellative, i.e. Bool $\not\subseteq$ Can$_c$. We note that Bool $\cap$ Can = O.

Corollary 8.2. *The variety Can contains uncountably many subvarieties of representable cancellative wPEMV-algebras. Consequently, the variety of representable wPEMV-algebras contains uncountably many subvarieties of representable wPEMV-algebras.*

Proof. Due to [8, Thm 61.24], the variety of representable ℓ-groups contains uncountably many subvarieties of representable ℓ-groups. Applying the proof of Theorem 8.1, we conclude the results. □

Lemma 8.3. *Let Wcom be the class of weakly commutative wPEMV-algebras. Then Wcom is a variety. In addition, it contains uncountably many subvarieties.*

Proof. First we note that a wPEMV-algebra with top element is weakly commutative iff $x^- = x^\sim$ for all $x \in M$. Indeed, if M is weakly commutative, then from $x^- + x = 1$, we conclude $x + x^-$ is defined in M, so that $x^- \leq x^\sim$. And similarly, $x + x^\sim = 1$ entails $x^\sim + x$ is defined so that $x^\sim \leq x^-$. Conversely, suppose $x^- = x^\sim$ for each $x \in M$. Let $x + y$ be defined in M. Then $x \leq y^- = y^\sim$, so that $y + x$ is defined and vice versa.

Let M_1 be a subalgebra of $M \in$ Wcom. Let for $x, y \in M_1$, $x + y$ be defined in M_1. Then $x + y$ is defined in M and consequently, $y + x$ is defined in M. Since $y + x = y \oplus x =: c \in M_1$, we have $y \leq c \ominus x \in M_1$, and $y + x$ is defined in M_1 giving $M_1 \in$ Wcom.

If $(M_t)_{t \in T}$ is a system of weakly commutative wPEMV-algebras, then the direct product $\prod_{t \in T} M_t$ is evidently also weakly commutative.

Let M be weakly commutative and let $\phi : M \to N$ be a surjective homomorphism of wPEMV-algebras. If M has a top element 1, then $\phi(1)$ is a top element of N. Since M is weakly commutative, we have $x^\sim = x^-$ for each $x \in M$, so that $\phi(1) \ominus \phi(x) = \phi(1 \ominus x) = \phi(x \odot 1) = \phi(x) \odot \phi(1)$ establishing that N is also weakly commutative.

Assume that M has no top element. We have two cases. Case (i): N is with top element. Then there is $u \in M$ such that $\phi(u)$ is a top element of N. For each $x \in [0, u]$, we have $\lambda_u(x) + x = u$, so that $x + \lambda_u(x)$ is defined in M. Due to Lemma 4.4, for each $c \geq u$ we have $\lambda_u(x) \leq \rho_c(x)$, which gives $\phi(u) \ominus \phi(x) \leq \phi(x) \odot \phi(c) \leq$

$\phi(x) \ominus \phi(u)$. In a similar way, from $x + \rho_u(x) = u$, we derive $\phi(x) \ominus \phi(u) \le \phi(u) \ominus \phi(x)$. Now, let $x \in M$ be arbitrary. Then $\phi(x) = \phi(x \wedge u)$ which by the last reasoning entails $\phi(x)^- = \phi(x \wedge u)^- = \phi(x \wedge u)^\sim = \phi(x)^\sim$ and this implies $N \in \mathsf{Wcom}$.

Case (ii): N has no top element. Assume that for some $x, y \in M$, $\phi(x) + \phi(y)$ is defined in N. Then $\phi(x) + \phi(y) = \phi(x) \oplus \phi(y) = \phi(x \oplus y)$. Due to Corollary 7.6, M has a weakly commutative representing wPEMV-algebra with top element. Clearly $(x \wedge y^-) + y$ is defined in M, $(x \wedge y^-) + y = x \oplus y$, and $x \wedge y^- = (x \oplus y) \ominus y$. Since M is weakly commutative, the element $y + (x \wedge y^-) = y + (x \wedge y^\sim)$ is also defined in M and due to (8) of the proof of Theorem 7.3, we have $y + (x \wedge y^-) = y + ((x \oplus y) \ominus y) = y + (y \ominus (y \oplus x))$. Then $\phi(y) + \phi(x \wedge y^-)$ exists in N and $\phi(y) + \phi(x \wedge y^-) = \phi(y) + \phi((x \oplus y) \ominus y) = \phi(y) + ((\phi(x) \oplus \phi(y)) \ominus \phi(y)) = \phi(y) + ((\phi(y) + \phi(y)) \ominus \phi(y)) = \phi(y) + \phi(x) \in N$.

Summarizing all three items, we have Wcom is a subvariety of wPEMV-algebras.

Since $\mathsf{Can} \subseteq \mathsf{Wcom}$ and by Corollary 8.2, Can contains uncountably many subvarieties, and they are subvarieties also of Wcom. $\square$

In [21, Thm 3.21], it was shown that the variety of (commutative) wEMV-algebras is the least variety containing all EMV-algebras (= associated wEMV-algebras). In Corollary 6.8, this result was extended to the class of representable associated wPEMV-algebras. Now we extend this result for the class of weakly commutative wPEMV-algebras.

Corollary 8.4. *The variety* Wcom *of weakly commutative wPEMV-algebras is the least subvariety of wPEMV-algebras containing the class of associated weakly commutative wPEMV-algebras.*

Proof. Due to Lemma 8.3, the class of weakly commutative wPEMV-algebras is a variety and every weakly commutative wPEMV-algebra possesses a representing wPEMV-algebra with top element. The rest of the proof is the same as the one of Corollary 6.8. $\square$

We note that it can happen that if a wPEMV-algebra M without top element belongs to some subvariety, so it is true for its representing wPEMV-algebra N with top element. For example, due to Theorem 6.2, a wPEMV-algebra belongs M to the variety Repr iff its representing wPEMV-algebra N also belongs to Repr. In general, it can happen that M belongs to a variety V but $N \notin \mathsf{V}$. This is true for example for the wPEMV-algebra of the positive cone $G^+ \in \mathsf{Can}$ but $\Gamma_a(\mathbb{Z} \overrightarrow{\times} G, (1,0)) \notin \mathsf{Can}$; this is true for each ℓ-group G.

In Proposition 6.7 it was shown that if $M \in \mathsf{Repr}$, then also the interval wPEMV-algebra $M_a = ([0, a]; \vee, \wedge, \oplus_a, \ominus, \odot, 0)$ also belongs to the same variety for each $a \in M$. In the next proposition we show how this result holds for other subvarieties.

Proposition 8.5. *Suppose a wPEMV-algebra M belongs to a subvariety V of wPEMV-algebras. If $a \in \mathcal{I}(M)$, then the interval wPEMV-algebra $M_a = [0, a]$ also belongs to V.*

Proof. To show different techniques used in theory of wPEMV-algebras, we present two proofs.

(i) Define a mapping $f_a : M \to [0, a]$ by $f_a(x) = x \wedge a$, $x \in M$. We show that f_a is a homomorphism of wPEMV-algebras. Clearly $f_a(0) = 0$ and by Proposition 5.2, we have that f_a preserves $\oplus$. By Theorem 4.3, the algebra $([0, a]; \oplus_a, \lambda_a, \rho_a, 0, a)$ is a pseudo MV-algebra. According to Corollary 4.7, $([0, a]; +_a, 0, a)$ and $([0, b]; +_b, 0, b)$ are PEAs, where $b \geq a$. Take $x, y \in [0, b]$ and let $x +_b y$ be defined in $[0, b]$. We show that $(x \wedge a) +_a (y \wedge a)$ is defined in the PEA $([0, a]; +_a, 0, a)$ and

$$(x \wedge a) +_a (y \wedge a) = (x +_b y) \wedge a. \tag{$*$}$$

Since $b \oplus_b b = b$ and $a \oplus a = a = a \oplus_b a$, a and b are Boolean elements in the pseudo MV-algebra $[0, b]$, for them Proposition 5.3(ii) holds. Then $x \leq \lambda_b(y)$, so that

$$x \wedge a \leq \lambda_b(y) \wedge a \leq \lambda_b(y \wedge a) \wedge a = \lambda_a(y \wedge a),$$

which implies $(x \wedge a) +_a (y \wedge a)$ is defined in the PEA $[0, a]$. Moreover, $(x \oplus y) = (x +_b y) \wedge a \geq (x \wedge a) +_b (y \wedge a) = (x \wedge a) +_a (y \wedge a) = (x \wedge a) \oplus (y \wedge a)$, which proves $(*)$. Equation $(*)$ implies also if $u \leq v \leq b$, then $(v \setminus_b u) \wedge a = (v \wedge a) \setminus_a (u \wedge a)$ and $(u /_b v) \wedge a = (u \wedge a) /_a (v \wedge a)$. Therefore, the restriction of f_a onto the PEA $[0, b]$ is a surjective homomorphism of pseudo effect algebras.

Now, let $x, y \in M$ and $x, y \leq b$. Then

$$(x \ominus y) \wedge a = (x \ominus (x \wedge y)) \wedge a =$$
$$= (x \setminus_b (x \wedge y)) \wedge a = (x \wedge a) \setminus_a (x \wedge y \wedge a)$$
$$= (x \wedge a) \ominus (y \wedge a).$$

In a dual way, we show also $(x \oslash y) \wedge a = (x \wedge a) \oslash (y \wedge a)$. Therefore, the mapping $f_a : M \to [0, a]$ is a surjective homomorphism of wPEMV-algebras, consequently, the wPEMV-algebra $M_a \in \mathsf{V}$.

(ii) Since a is an idempotent, the interval $[0, a]$ is a normal ideal of M which is closed under $\vee, \wedge, \oplus, \ominus, \oslash$, and 0 so it is a subalgebra of M which implies the wPEMV-algebra $M_a \in \mathsf{V}$. $\square$

On the other hand, if we take the positive cone G^+ of a non-zero ℓ-group G, then $G^+ \in \mathsf{Can}$ but $M_a \notin \mathsf{Can}$ for each $a \in G \setminus \{0\}$.

In addition, it is important to underline that if $a \in M$ is not idempotent, then the mapping $f_a : M \to [0, a]$ defined by $f_a(x) = x \wedge a$, $x \in M$, is not necessarily a homomorphism.

It is worthy of notifying that if a wPEMV-algebra M without top element admits a representing wPEMV-algebra N with top element, then on N we have the associative binary operation $\odot$ that is defined by $x \odot y = (y^- \oplus x^-)^\sim$, $x, y \in N$. If $x, y \in M$, then $x \odot y \in M$. Hence, for each $x \in N$, and any integer $n \geq 0$, we define

$$x^0 := 1, \quad x^n := x^{n-1} \odot x, \quad \text{if } n \geq 1.$$

In what follows, we describe the variety consisting of all representing wPEMV-algebras of Can.

Let Perf denote the class of wPEMV-algebras of the form $(\Gamma_a(\mathbb{Z} \vec{\times} G, (1, 0)); \vee, \wedge, \oplus, \ominus, \sim, 0)$, where G is any ℓ-group. Then $\Gamma_a(\mathbb{Z} \vec{\times} G, (1, 0))$ is a weakly commutative and wPEMV-algebra with top element representing the cancellative wPEMV-algebra of the positive cone G^+. The class Perf is not a variety because it is not closed under the direct product. Indeed, the wPEMV-algebra $\Gamma_a(\mathbb{Z} \vec{\times} \mathbb{Z}, (1, 0)) \times \Gamma_a(\mathbb{Z} \vec{\times} \mathbb{Z}, (1, 0))$ does not belong to Perf. Such a class of MV-algebras, called perfect MV-algebras, was studied in [10].

Theorem 8.6. *Let $V_0(\text{Perf})$ be the subvariety generated by class Perf and let BP be the variety of weakly commutative wPEMV-algebras that satisfy the equation*

$$(2.x)^2 = 2.x^2. \tag{8.1}$$

Then $V_0(\text{Perf}) = \text{BP}$ and Can is a proper subvariety of $V_0(\text{Perf})$ and if $G = \text{Aut}(\mathbb{R})^+$, then $\Gamma_a(\mathbb{Z} \vec{\times} G, (1, 0))$ is a generator of $V_0(\text{Perf})$.

Proof. Due to Lemma 8.3, the class of weakly commutative wPEMV-algebras is a variety. It is easy to verify that every cancellative wPEMV-algebra satisfies the equation (8.1). Then $\text{Can} \subsetneq V_0(\text{Perf})$ because $\Gamma_a(\mathbb{Z} \vec{\times} G, (1, 0)) \in V_0(\text{Perf}) \setminus \text{Can}$ for every ℓ-group G. We note that by Theorem 8.1, the positive cone $\text{Aut}(\mathbb{R})^+$ generates the variety Can, hence, $\Gamma_a(\mathbb{Z} \vec{\times} \text{Aut}(\mathbb{R})^+, (1, 0))$ generates the variety $V_0(\text{Perf})$.

It is also easy to verify that every perfect wPEMV-algebra $\Gamma_a(\mathbb{Z} \vec{\times} G, (1, 0))$ is weakly commutative and satisfies (8.1), so that $V_0(\text{Perf}) \subseteq \text{BP}$.

On the other side, if $M \in \text{BP}$ has a top element, then it is equivalent to a pseudo MV-algebra, and by [9, Thm 6.6], this pseudo MV-algebra belongs to the variety generated by all pseudo MV-algebras $\Gamma(\mathbb{Z} \vec{\times} G, (1, 0))$, $G \in \mathcal{LG}$. Whence, M belongs to $V_0(\text{Perf})$, and in addition, every wPEMV-subalgebra of M also belongs to $V_0(\text{Perf})$. Now, let $M \in \text{BP}$ do not have a top element. Since M is weakly commutative, by Corollary 7.6, M possesses a representing weakly commutative wPEMV-algebra N with top element. Without loss of generality, we can assume that $M \subset N$. Then every element $x \in M$ satisfies (8.1). If $x \in N \setminus M$, then $x = x_0^\sim$ for some $x_0 \in M$. It is easy to verify that x also satisfies equation (8.1), i.e. $N \in \mathsf{BP}$ and both N and M also belong to $\mathsf{V}_0(\mathsf{Perf})$. In other words, $\mathsf{BP} \subseteq \mathsf{V}_0(\mathsf{Perf})$, and finally $\mathsf{BP} = \mathsf{V}_0(\mathsf{Perf})$. $\square$

In what follows, we will interested in normal-valued wPEMV-algebras. Therefore, we start with presenting the following result.

Proposition 8.7. *Let I be an ideal of a wPEMV-algebra M and let $g \in M \setminus I$. Then there is an ideal V of M containing I and not containing g that is maximal with respect to this property. Moreover, V is prime.*

Proof. Applying the Zorn lemma, there is an ideal V of M maximal with respect to the property containing I and not containing g. Let A and B be two ideals of M which properly contain V. Then they contain also the element g. Therefore, $V \subsetneq A \cap B$ which by (vii) of Lemma 5.4 guarantees that V is prime. $\square$

If $I = \{0\}$, every ideal V from Proposition 8.7 is said to be a *value* of $g > 0$. The ideal $V^* = V \vee I_0(g)$ is said to be a *cover* of V. A wPEMV-algebra M is said to be *normal-valued* if every value V of $g > 0$ is normal in its cover (i.e. $g \oplus V = V \oplus g$ for each $g \in V^*$). In [13, Thm 6.8], the class of normal-valued pseudo MV-algebras forms a variety. Its equational base is

$$(x \oplus y) \wedge (2.y \oplus 2.x) = x \oplus y. \tag{8.2}$$

This notion corresponds to the notion of normal-valued ℓ-group. An ℓ-group G is normal-valued iff it satisfies the equation $g + h \leq 2h + 2g$ for all $g, h \in G^+$, [8, Thm 41.1], and the variety of normal-valued ℓ-groups is the biggest proper subvariety of ℓ-groups, see [8, Cor 58.12]. Nevertheless that a pseudo MV-algebra $M = \Gamma(G, u)$ is normal-valued iff G is normal-valued, [13, Prop 6.2], the variety of normal-valued pseudo MV-algebras is not the biggest proper subvariety of the variety of MV-algebras because, it is a proper subvariety of the variety of pseudo MV-algebras where every maximal ideal is normal, see [15, Prop 6.2]; it does not have a parallel with ℓ-groups. Therefore, we can formulate the following corollary.

Corollary 8.8. *The class of cancellative wPEMV-algebras satisfying equation (8.2) is the biggest proper subvariety of the variety* **Can**.

In the next result we show that the class of normal-valued wPEMV-algebras forms a variety.

Theorem 8.9. *The class* NV *of normal-valued wPEMV-algebras forms a variety and the variety of representable wPEMV-algebras is a proper subvariety of* NV. *In addition, the class of cancellative and normal-valued wPEMV-algebras is the biggest proper subvariety of the variety* Can.

Proof. We show that the class NV is closed under forming subalgebras, homomorphic images, and direct products.

(1) *If M_1 is a subalgebra of $M \in$ NV, then $M_1 \in$ NV.*

Let V_1 and V_1^* be a value and its cover, respectively, of an element $g \in M_1 \setminus \{0\}$. Let $\langle V_1 \rangle$ be the ideal of M generated by V_1. Then $g \notin \langle V_1 \rangle$. Let V be an ideal of M maximal with respect to the condition V contains $\langle V_1 \rangle$ and does not contain g. In addition, let V^* be its cover in M. Then $M_1 \cap V = V_1$ and $M_1 \cap V^* = V_1^*$. If $x, y \in V_1^*$, then the normality of V in V^* implies that $x \obar y \in V_1$ iff $y \obar x \in V_1$.

(2) *If $M_1 \in$ NV and $\psi : M_1 \to M_2$ is a surjective homomorphism, then $M_2 \in$ NV.*

Let V_2 be a value of $g \in M_2 \setminus \{0\}$ and V_2^* be its cover. Let $V_1 = \psi^{-1}(V_2)$ and $V_1' = \psi^{-1}(V_2^*)$. It is straightforward to verify that V_1 and V_1' are ideals of M_1. Let $x_0 \in \psi^{-1}(\{g\})$. Then $x_0 \in V_1' \setminus V_1$. Let V_0 be the ideal of M_1 generated by V_1 and x_0. Then $\psi(V_0)$ is an ideal of M_2 containing V_2 and contained in V_2^*, which shows $\psi(V_0) = V_2^*$.

Now, we show that $\psi^{-1}(V_2^*) = V_0$. Clearly, $V_0 \subseteq \psi^{-1}(V_2^*)$. On the other hand, let $x \in \psi^{-1}(V_2^*)$. Then $\psi(x) \leq v_1 \oplus g \oplus \cdots \oplus v_n \oplus g =: z$ for some $v_1, \ldots, v_n \in V_2$ and some $n \geq 1$. There are $u_1, \ldots, u_n \in V_1$ such that $\psi(u_i) = v_i$ for each $i = 1, \ldots, n$ and the element $w = u_1 \oplus x_0 \oplus \cdots \oplus u_n \oplus x_0$ belongs to V_0. Therefore,

$$\psi(x \ominus w) = \psi(x) \ominus \psi(w) = \psi(x) \ominus z = 0,$$

which implies $x \ominus w \in \psi^{-1}(\{0\}) \subseteq \psi^{-1}(V_2) = V_1 \subseteq V_0$. Then due to (W4), $x \vee w = (x \ominus w) \oplus w \in V_0$ and finally $x \in V_0$, which finishes the proof that $V_0 = \psi^{-1}(V_2^*) = V_1'$. This yields V_2 is a value of x_0 and $V_1^* = V_1' = V_0$ is its cover. Now, if $\psi(x), \psi(y) \in V_1^*$, we have $\psi(x) \ominus \psi(y) = \psi(x \ominus y)$ and $\psi(y) \obar \psi(x) = \psi(y \obar x)$, and the normality of V_1 in V_1^* entails the normality of V_2 in V_2^*, establishing $M_2 \in$ NV.

(3) *If M is a direct product of normal-valued wPEMV-algebras, then $M \in$ NV.*

Let $M = \prod_i M_i$, where every M_i belongs to NV. Let V be a value of some $g = (g_i)_i > 0$ in M and let V^* be the cover of V. Define $V_i := \pi_i(V)$ and $V_i^* = \pi_i(V^*)$ for each i, where π_i is the i-th projection from M onto M_i. Then either $V_i \subsetneq V_i^*$ or $V_i = V_i^*$.

In the first case, fix i_0, then $g_{i_0} \in V_{i_0}^* \setminus V_{i_0}$ and we assert V_{i_0} is a value of g_{i_0}. Take an ideal U_{i_0} of M_{i_0} containing V_{i_0} and not containing g_{i_0}. Then $U = \prod_i A_i$, where $A_i = U_{i_0}$ if $i = i_0$ and $A_i = \{0_i\}$ otherwise, is an ideal of M containing V and it does not contain g. The ideal $V \vee U$ has to contain g, so that $g \leq (v_i^1)_i \oplus (u_i^1)_i \oplus \cdots \oplus (v_i^n)_i \oplus (u_i^n)_i$, where $(v_i^k)_i \in V$ and $(u_i^k)_i \in U$ for $k = 1, \ldots, n$. Then $g_{i_0} \leq v_{i_0}^1 \oplus u_{i_0}^1 \oplus \cdots \oplus v_{i_0}^n \oplus u_{i_0}^n$. But all $v_{i_0}^k, u_{i_0}^k \in U_{i_0}$ which yields a contradiction $g_{i_0} \in U_{i_0}$. Therefore, V_{i_0} is a value of g_{i_0}.

Now, we show that $V_{i_0}^*$ is a cover of V_{i_0}. Let $y_{i_0} \in V_{i_0}^* \setminus V_{i_0}$ and let y be an arbitrary element of $V^* \setminus V$ whose the i_0-th coordinate is y_{i_0}. There are $(v_i^1)_i, \ldots, (v_i^j)_i \in V$ such that $y \leq (v_i^1)_i \oplus g \oplus \cdots \oplus (v_i^j)_i \oplus g$, which implies $y_{i_0} \leq v_{i_0}^1 \oplus g_{i_0} \oplus \cdots \oplus v_{i_0}^j \oplus g_{i_0}$, where $v_{i_0}^1, \ldots, v_{i_0}^j \in V_{i_0}$. This proves that $V_{i_0}^*$ is a cover of V_{i_0} in M_{i_0}.

Finally, let $x = (x_i)_i$ and $y = (y_i)_i$ be elements of V^*. In the first situation when $V_i \subsetneq V_i^*$, we have $x_i, y_i \in V_i^*$ and $x_i \ominus y_i \in V_i$ iff $y_i \odot x_i \in V_i$. If $V_i = V_i^*$, V_i is trivially normal in V_i^*. Therefore, V_i is normal in V_i^* for each i. Consequently, $M \in \mathsf{NV}$.

Now let M be any linearly ordered wPEMV-algebra. Due to Theorem 5.8, either M is with top element or M is cancellative. In the first case, Theorem 6.2 says M possesses a representable and representing wPEMV-algebra N with top element. Therefore, N can be viewed as a representable pseudo MV-algebra and according to [13, Thm 6.11], N is normal-valued, so that $M \in \mathsf{NV}$. If M is cancellative, then $M \cong G^+$ for some ℓ-group G and $N = \Gamma_a(\mathbb{Z} \overset{\rightarrow}{\times} G, (1,0))$ is its representing and representable wPEMV-algebra with top element. Also in this case N as a pseudo MV-algebra is a normal-valued pseudo MV-algebra, consequently $M \in \mathsf{NV}$. In either case $\mathsf{Repr} \subseteq \mathsf{NV}$. Since there is a normal-valued pseudo MV-algebra that is not representable, [13, Ex 2.2], Repr is a proper subvariety of NV.

By a way, the class of cancellative normal-valued wPEMV-algebras is the biggest proper subvariety of the variety Can. □

We note that a special class of normal-valued residuated lattices was studied in [32] and due to [5, Thm 6.6], a basic pseudo hoop is normal-valued iff it satisfies (8.2) and a sequence of additional identities. For pseudo MV-algebras the only identity (8.2) is equivalent to be normal-valued, as it was already said.

Question 8.10. Is (8.2) an equational base for the variety NV?

Question 8.11. The variety of normal-valued ℓ-groups is the biggest proper subvariety of the variety of ℓ-groups. However, the variety of normal-valued pseudo MV-algebras is not the biggest proper subvariety of the variety of pseudo MV-algebras, it is a proper subvariety of the variety of pseudo MV-algebras, where each maximal ideal is normal, see [15, Prop 6.2]. Question: Is the subvariety of normal-valued wPEMV-algebras the biggest proper subvariety of the variety wPEMV?

9 Conclusion

In the paper, we have introduced weak pseudo EMV-algebras which are a non-commutative generalization of MV-algebras, generalized Boolean algebras, EMV-algebras, pseudo MV-algebras and wEMV-algebras. The paper is divided into two parts.

Part I: We studied the basic properties of weak pseudo EMV-algebras as algebras with bottom element but top element is not assumed a priori. We presented important examples of wPEMV-algebras as cancellative ones and associated wPEMV-algebras. Using Bosbach's [3] notion of a semiclan and deriving a partial addition +, we have shown that every wPEMV-algebra $(M; \vee, \wedge, \oplus, \ominus, \oslash, 0)$ can be embedded into the positive cone of some ℓ-group preserving all operations besides $\oplus$, Theorem 4.6. In Theorem 4.3, we have established that every interval $[0, c]$ in a wPEMV-algebra can be converted into a pseudo MV-algebra $([0, c]; \oplus_c, \lambda_c, \rho_c, 0, c)$, where $x \oplus_c y = (x \oplus y) \wedge c$, see Theorem 4.3. Moreover, Proposition 4.9 shows that wPEMV-algebras can be studied also in an equivalent way as integral GMV-algebras in the sense of [2, 25], a special class of residuated lattices. Using it, some results can be established also in this way which were established also for GMV-algebras, see [1, 2, 25].

We investigated a main question when a wPEMV-algebra M without top element can be embedded into a wPEMV-algebra N with top element as a maximal and normal ideal of N such that every element outside of the image of M in N is a complement of some element from the image of M. N is said to be representing the PEMV-algebra M. A sufficient condition was given in Theorem 5.14. First such a question was exhibited in [7] for generalized Boolean algebras and for EMV-algebras, pseudo EMV-algebras and wEMV-algebras in [18, 20, 21].

Part II: One of classes, where each wPEMV-algebra admits a PEMV-algebra representing it, is the class of representable wPEMV-algebras. We have showed that such a wPEMV-algebra representing it is also a representable wPEMV-algebra, Theorem 6.2. The general solution is presented in the Basic Representation Theorem 7.9.

We introduced unitizing left and right automorphisms of wPEMV-algebras and we showed that the existence of a left (right) unitizing automorphism is a necessary and sufficient condition in order that a wPEMV-algebra does possess a representing wPEMV-algebra with top element, Theorem 7.3 and Theorem 7.5. We showed that the class of cancellative or weakly commutative wPEMV-algebras admits such unitizing automorphisms. Using the Basic Representation Theorem 7.9, we have that every wPEMV-algebra without top element admits unitizing automorphisms.

We have studied some interesting wPEMV-algebras which form a variety: The variety of cancellative wPEMV-algebras, Theorem 8.1, the variety of perfect wPEMV-algebras, Theorem 8.6, with equational base $(2.x)^2 = 2.x^2$, and the variety of normal-valued wPEMV-algebras, Theorem 8.9.

The paper is accomplished with some open questions.

The presented results show how big can be a variety of different generalizations of MV-algebras, or equivalently the variety of integral GMV-algebras. In future, we hope to extend our knowledge on these algebras. As we have said, we know that some results for wPEMV-algebras can be described also in the language of integral GMV-algebras and one of our the basic results applied to integral GMV-algebras without least element says that such an integral GMV-algebra can be embedded into a bounded integral GMV-algebra as a maximal and normal filter.

AcknowledgementS

The authors are very indebted to anonymous referees for their careful reading, observations and suggestions which helped us to improve the presentation of the paper. In particular, we are indebted for a hint to look at wPEMV-algebras also as at integral GMV-algebras.

References

[1] P. Bahls, J. Cole, N. Galatos, P. Jipsen, C. Tsinakis, *Cancellative residuated lattices*, Algebra Universalis **50** (2003), 83–106.

[2] K. Blount, C. Tsinakis, *The structure of residuated lattices*, Inter. J. Algebra **13** (2003), 437–461.

[3] B. Bosbach, *Concerning semiclans*, Arch. Math. **37** (1981), 316–324.

[4] B. Bosbach, *Concerning cone algebras*, Algebra Universalis **15** (1982), 58–66.

[5] M. Botur, A. Dvurečenskij, T. Kowalski, *On normal-valued basic pseudo hoops*, Soft Computing **16** (2012), 635–644. DOI: 10.1007/s00500-011-0763-7

[6] S. Burris and H.P. Sankappanavar, *A Course in Universal Algebra*, Springer-Verlag, New York, Heidelberg, 1981.

[7] P. Conrad, M.R. Darnel, *Generalized Boolean algebras in lattice-ordered groups*, Order **14** (1998), 295–319.

[8] M.R. Darnel, *Theory of Lattice-Ordered Groups*, Marcel Dekker, Inc., New York, Basel, Hong Kong, 1995.

[9] A. Di Nola, A. Dvurečenskij, C. Tsinakis, *Perfect GMV-algebras*, Comm. Algebra **36** (2008), 1221–1249.

[10] A. Di Nola, A. Lettieri, *Perfect MV-algebras are categorically equivalent to Abelian ℓ-groups*, Studia Logica **53** (1994), 417–432.

[11] A. Di Nola, A. Lettieri, *Equational characterization of all varieties of MV-algebras*, J. Algebra **221** (1999), 463–474.

[12] A. Dvurečenskij, *On pseudo MV-algebras*, Soft Computing **5** (2001), 347–354.

[13] A. Dvurečenskij, *States on pseudo MV-algebras*, Studia Logica **68** (2001), 301–327.

[14] A. Dvurečenskij, *Pseudo MV-algebras are intervals in ℓ-groups*, J. Austral. Math. Soc. **72** (2002), 427–445.

[15] A. Dvurečenskij, W.C. Holland, *Top varieties of generalized MV-algebras and unital lattice-ordered groups*, Comm. Algebra **35** (2007), 3370–3390.

[16] A. Dvurečenskij, T. Vetterlein, *Pseudoeffect algebras. I. Basic properties*, Inter. J. Theor. Phys. **40** (2001), 685–701.

[17] A. Dvurečenskij, T. Vetterlein, *Pseudoeffect algebras. II. Group representation*, Inter. J. Theor. Phys. **40** (2001), 703–726.

[18] A. Dvurečenskij, O. Zahiri, *On EMV-algebras*, Fuzzy Sets and Systems **373** (2019), 116–148.

[19] A. Dvurečenskij, O. Zahiri, *Pseudo EMV-algebras. I. Basic Properties*, Journal of Applied Logics–IFCoLog Journal of Logics and their Applications **6** (2019), 1285–1327.

[20] A. Dvurečenskij, O. Zahiri, *Pseudo EMV-algebras. II. Representation and States*, Journal of Applied Logics–IFCoLog Journal of Logics and their Applications **6** (2019), 1329–1372.

[21] A. Dvurečenskij, O. Zahiri, *A variety containing EMV-algebras and Pierce sheaves of EMV-algebras*, Fuzzy Sets and Systems **418** (2021), 101–125. https://doi.org/10.1016/j.fss.2020.09.011

[22] A. Dvurečenskij, O. Zahiri, *Weak pseudo EMV-algebras. I. Basic properties*, Journal of Applied Logics–IFCoLog Journal of Logics and their Applications, **8** (2021), 2365–2399.

[23] D.J. Foulis, S. Pulmannová, *Unitizing a generalized pseudo effect algebra*, Order **32** (2015), 189–204.

[24] L. Fuchs, *Partially Ordered Algebraic Systems*, Pergamon Press, Oxford-New York, 1963.

[25] N. Galatos, C. Tsinakis, *Generalized MV-algebras*, J. Algebra **283** (2005), 254–291.

[26] G. Georgescu and A. Iorgulescu, *Pseudo MV-algebras*, Multiple-Valued Logics **6** (2001), 193–215.

[27] A.M.W. Glass, *Partially Ordered Groups*, World Scientific, Singapore, New Yersey, London, Hong Kong, 1999.

[28] J. Kühr, *Pseudo BL-algebras and DRℓ-monoids*, Math. Bohemica **128** (2003), 199–208.

[29] W. Rump, Y. Yang, *A note on Bosbach's cone algebras*, Studia Logica **98** (2011), 375–386.

[30] W. Rump, Y. Yang, *Pseudo MV-algebras as L-algebras*, J. Mult.-Valued Logic Soft Computing **19** (2012), 621–632.

[31] N.V. Subrahmanyam, *Bricks and pseudo MV-algebras are equivalent*, Math. Slovaca **58** (2008), 131–142.

[32] C. Tsinakis, A. Ledda, F. Paoli, *The Archimedean property: New perspectives*, Talk at BLAST 2021, June 9-13, 2021, Las Cruces, N. Mexico. https://math.nmsu.edu/blast-2021/slides/Tsinakis.pdf

Received 25 May 2021

A 2 Set-up Routley Semantics for the 4-valued Logic PŁ4

Gemma Robles
Dpto. de Psicología, Sociología y Filosofía, Universidad de León.
Campus de Vegazana, s/n, 24071, León, Spain
gemma.robles@unileon.es

José M. Méndez
Edificio FES, Universidad de Salamanca.
Campus Unamuno, 37007, Salamanca, Spain
sefus@usal.es

Abstract

The paraconsistent and paracomplete 4-valued logic PŁ4 is endowed with a 2 set-up Routley semantics.

Keywords: Paraconsistent logics; paracomplete logics; 4-valued logics; modal 4-valued logics; Routley semantics; 2 set-up Routley semantics.

1 Introduction

The paraconsistent and paracomplete 4-valued logic PŁ4 is introduced in [8]. PŁ4 is characterized by a modification of the matrix determining Łukasiewicz's 4-valued modal logic Ł (cf. [6], [7]). In [8], it is shown that PŁ4 is a strong and rich logic that is free from the Łukasiewicz-type modal paradoxes afflicting the system Ł (cf. [9] and references therein). In [8], it is also proved that PŁ4 can be given a simple two-valued Belnap-Dunn semantics.

In [5], Section 1, it is noted that PŁ4 is equivalent to De and Omori's logic BD$_+$ (cf. [4], Zaitsev's paraconsistent logic FDEP (cf. [17]) and Béziau's four-valued modal logic PM4N (cf. [2]). These four equivalent logics were obtained

---

This work is supported by MCIN/AEI/ 10.13039/501100011033 [Grant PID2020-116502GB-I00]. We sincerely thank an anonymous referee of the JAL for her(his) comments and suggestions on a previous draft of this paper.

from different motivations, which we think shows that they do not correspond to an artificial construct, but that they are four different versions of a natural logic.

The aim of this paper is to present still another perspective on PŁ4 by interpreting the negation characteristic of this logic by using the Routley operator or Routley star (cf. [14], [15], [16] and references in the last item). (It is worth remarking that there are antecedents of the Routley operator in the classical Polish logical school, particularly, in the works of Białynicki-Birula and Rasiowa —cf. [1], §48.2]). More specifically, the aim of the present paper is to endow PŁ4 with a very simple *2 set-up Routley semantics*.

2 set-up Routley-Meyer semantics (RM-semantics) is defined in [3], where the logics BN4, RM3 and Łukasiewicz's 3-valued logic Ł3 are interpreted with a 2 set-up RM-semantics. In addition, in [11], the logic E4 is also given a 2 set-up RM-semantics. In the papers just quoted, models are based upon structures of the type $(K, R, *)$, where K is the 2 set-up set, $*$ is the Routley star and R is a ternary relation defined on K. However, in the present paper, the ternary relation is dropped, that is, the structures of interest are of the type $(K, *)$, hence the label '2 set-up Routley semantics', instead of '2 set-up Routley-Meyer semantics[1]'.

The paper is organized as follows. In §2, the logic PŁ4 is recalled. In §3, PŁ4 is given a 2 set-up Routley semantics and the soundness theorem is proved. In §4, completeness of PŁ4 w.r.t. the semantics defined in §3 is proved. Finally, in §5, we note some brief remarks on possible future work to be done on 2 set-up Routley semantics. We have added an appendix on some of the propositional connectives definable in PŁ4.

2 The logic PŁ4

In this section the logic PŁ4 defined in [8] is recalled.

The propositional language consists of a denumerable set of propositional variables $p_0, p_1, ..., p_n, ...$, and the following connectives: $\to$ (conditional) and $\neg$ (negation). The set of wffs is defined in the customary way. A, B, C, etc. are metalinguistic variables. PŁ4 is formulated as a Hilbert-type axiomatic system, the notions of 'theorem' and 'proof from a set of premises' being understood in the standard way.

---

[1]We remark that in [10], §6.1, the Routley star is used to interpret classical negation in order to provide a semantics for the logic BD$_+$ (cf. [4]) understood as an extension of Anderson and Belnap's *First degree entailment logic*, FDE (cf. [1] and references therein) with classical negation. However, this semantics for BD$_+$ is neither a set-up Routley semantics nor a semantics for PŁ4, which is defined with a De Morgan, not a classical, negation. (Nevertheless, classical negation is definable in PŁ4 —cf. [8], Proposition 7.8.)

Definition 1 (The logic PŁ4). *The logic PŁ4 can be axiomatized as follows.*
Axioms:

$$A1.\ A \to (B \to A)$$
$$A2.\ [A \to (B \to C)] \to [(A \to B) \to (A \to C)]$$
$$A3.\ [(A \to B) \to A] \to A$$
$$A4.\ A \to \neg\neg A$$
$$A5.\ \neg\neg A \to A$$
$$A6.\ \neg(A \to B) \to (\neg A \to C)$$
$$A7.\ \neg(A \to B) \to \neg B$$
$$A8.\ \neg B \to [[\neg A \to \neg(A \to B)] \to \neg(A \to B)]$$

Rule of inference:

Modus Ponens (MP). $A, A \to B \Rightarrow B$ *(if A and $A \to B$, then B)*

Definition 2 (The matrix MPŁ4). *The propositional language consists of the connectives $\to$ and $\neg$. The matrix MPŁ4 is the structure $(\mathcal{V}, D, \mathbf{F})$ where (1) $\mathcal{V}$ is $\{0, 1, 2, 3\}$ and is partially ordered as shown in the following lattice*

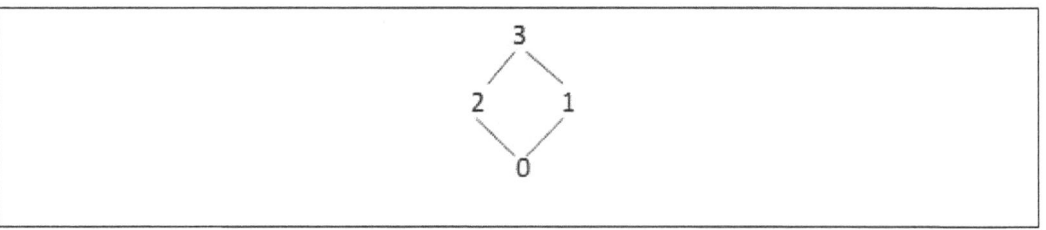

(2) $D = \{3\}$; $\mathbf{F} = \{f_\to, f_\neg\}$ where $f_\to$ and $f_\neg$ are defined according to the following truth-tables:

$\to$	0	1	2	3	$\neg$
0	3	3	3	3	3
1	2	3	2	3	1
2	1	1	3	3	2
3	0	1	2	3	0

In [8], it is proved that PŁ4 is determined by the degree of truth-preserving consequence relation defined on the ordered set of values of MPŁ4.

Remark 1 (Some theorems of PŁ4). *The following theorems of PŁ4 will be used in the sequel: (T1) $A \to A$; (T2) $(A \to C) \to [(B \to C) \to [(A \to B) \to B] \to C]]$; (T3) $\neg B \to [\neg A \to \neg[(A \to B) \to B]]$. (In the appendix to the paper, we have remarked some connectives definable in PŁ4, as well as some of its conspicuous theorems.)*

3 A 2 set-up Routley semantics

In this section, PŁ4 is endowed with a very simple 2 set-up Routley semantics. We begin by defining the concept of a model and related notions.

Definition 3 (2 set-up PŁ4-models). *Let $*$ be an involutive operation defined on the set K, that is, for any $x \in K$, $x = x^{**}$, and let K be the two-element set $\{0, 0^*\}$. A 2 set-up Routley PŁ4-model (2PŁ4-model, for short) is a structure $(K, *, \vDash)$ where $\vDash$ is a (valuation) relation from K to the set of all wffs such that the following conditions (clauses) are satisfied for every propositional variable p, wffs A, B and $a \in K$:*

(i). $a \vDash p$ or $a \nvDash p$

(ii). $a \vDash A \to B$ iff $a \nvDash A$ or $a \vDash B$

(iii). $a \vDash \neg A$ iff $a^* \nvDash A$

Definition 4 (2PŁ4-consequence, 2PŁ4-validity). *For a non-empty set of wffs Γ and wff A, $\Gamma \vDash_M A$ (A is a consequence of Γ in the 2PŁ4-model M) iff for all $a \in K$ in M, $a \vDash A$ whenever $a \vDash \Gamma$ ($a \vDash \Gamma$ iff $a \vDash B$ for all $B \in \Gamma$). Then, $\Gamma \vDash_{2PŁ4} A$ (A is a 2PŁ4-consequence of Γ) iff $\Gamma \vDash_M A$ in every 2PŁ4-model M.*

In particular, if $\Gamma = \emptyset$, $\vDash_M A$ (A is true in M) iff $a \vDash A$ for all $a \in K$ in M. And $\vDash_{2PŁ4} A$ (A is 2PŁ4-valid) iff $\vDash_M A$ in every 2PŁ4-model.

Next, we prove the soundness theorem.

Theorem 1 (Soundness of PŁ4). *For any set of wffs Γ and wff A, if $\Gamma \vdash_{PŁ4} A$, then $\Gamma \vDash_{2PŁ4} A$.*

Proof. If $A \in \Gamma$ or A has been derived by MP, the proof is trivial; A1-A3 are immediate by clause (ii) in Definition 3 and A4, A5 are easy by the same clause and involutiveness of $*$. So, let us prove the 2PŁ4-validity of A6, A7 and A8.

A6, $\neg(A \to B) \to (\neg A \to C)$, is 2PŁ4-valid: Let M be an arbitrary 2PŁ4-model where $a \in K$ and A, B, C wffs such that (1) $a \vDash \neg(A \to B)$ (i.e., $a^* \nvDash A \to B$) but (2) $a \nvDash \neg A \to C$. Then, we have (3) $a \vDash \neg A$ (i.e., $a^* \nvDash A$) and (4) $a \nvDash C$. By 1, we get (5) $a^* \vDash A$ and (6) $a^* \nvDash B$. But 3 and 5 contradict each other.

A7, $\neg(A \to B) \to \neg B$, is 2PŁ4-valid: Let M be an arbitrary 2PŁ4-model where $a \in K$ and A, B wffs such that (1) $a \vDash \neg(A \to B)$ (i.e., $a^* \nvDash A \to B$) but (2) $a \nvDash \neg B$ (i.e., $a^* \vDash B$). By 1, we get (3) $a^* \vDash A$ and (4) $a^* \nvDash B$. But 2 and 4 contradict each other.

A8, $\neg B \to [[\neg A \to \neg(A \to B)] \to \neg(A \to B)]$, is 2PŁ4-valid: Let M be an arbitrary 2PŁ4-model where $a \in K$ and A, B wffs such that (1) $a \vDash \neg B$ (i.e., $a^* \nvDash B$) but (2) $a \nvDash [\neg A \to \neg(A \to B)] \to \neg(A \to B)$. By 2, we have (3) $a \vDash \neg A \to \neg(A \to B)$ and (4) $a \nvDash \neg(A \to B)$. By 3 and 4, we get (5) $a \nvDash \neg A$ (i.e., $a^* \vDash A$); and by 4, we obtain (6) $a^* \vDash A \to B$, whence by 1, we have (7) $a^* \nvDash A$, contradicting 5.

If $\Gamma = \emptyset$, the proof is similar. □

Before proceeding to prove completeness, it may be interesting to present 2PŁ4-models falsifying some conspicuous classical tautologies. We consider two 2PŁ4-models where $0 \neq 0^*$ and distinct propositional variables p, q which are evaluated in each one of these models as follows (if either p or q is not evaluated in 0 —or in 0^*—, this means that the variable in question can arbitrarily be evaluated).

M1: $0 \vDash p, 0^* \nvDash p, 0 \nvDash q$.

M2: $0 \nvDash p, 0^* \vDash p$.

We have (1) $\neg A \to (A \to B)$ is falsified in M1: $0 \vDash \neg p$ but $0 \nvDash p \to q$; (2) $A \vee \neg A$ is falsified in M2: $0 \nvDash p, 0 \nvDash \neg p$; (3) $(\neg A \vee B) \to (A \to B)$ is falsified in M1: $0 \vDash \neg p \vee q$ but $0 \nvDash p \to q$. Now, in [8], it is proved that addition of any $\neg A \to (A \to B)$, $A \vee \neg A$ or $(\neg A \vee B) \to (A \to B)$ to PŁ4 results in a system in which all classical tautologies are derivable.

4 Completeness of PŁ4

We prove the completeness of PŁ4 w.r.t. the semantics displayed in the previous section by using a canonical model construction. That is, we show that if $\Gamma \nvdash_{PŁ4} A$, then there is a prime, a-consistent theory $\mathcal{T}$ (the notions are defined below) such that $\Gamma \subseteq \mathcal{T}$ and $A \notin \mathcal{T}$. This means that A is not a consequence of Γ from a canonical point of view to be defined, whence $\Gamma \nvDash_{2PŁ4} A$ follows. We begin by defining the notion of a theory and the classes of theories of interest in the present paper.

Definition 5 (PŁ4-theories). *A PŁ4-theory (theory, for short) is a set of formulas containing all theorems of PŁ4 and closed under Modus Ponens (MP). That is, a is a theory iff (1) if $\vdash_{PŁ4} A$, then $A \in a$; and (2) $B \in a$ whenever $A \to B \in a$ and $A \in a$.*

Definition 6 (Classes of theories). *Let a be a theory. We set (1) a is prime iff whenever $(A \to B) \to B \in a$, then $A \in a$ or $B \in a$; (2) a is trivial if it contains all wffs; (3) a is a-consistent ('consistent in an absolute sense') iff a is not trivial. Also, the following definitions will be used at the end of the section: (4) a is w-inconsistent ('inconsistent in a weak sense') iff $\neg A \in a$, A being some PŁ4-theorem; (5) a is w-consistent ('consistent in a weak sense') iff a is not w-inconsistent (cf. [13] and references therein on the notion of w-consistency).*

Corollary 1 (Closure under PŁ4-entailment). *Let a be a theory. Then a is closed under PŁ4-entailment. That is, if $\vdash_{PŁ4} A \to B$ and $A \in a$, then $B \in a$.*

Proof. It is immediate by the fact that a contains all PŁ4-theorems and it is closed under MP. □

In what follows, we proceed to the definition of the canonical 2PŁ4-model. We lean on the primeness lemma and the standard notion of "set of consequences of a set of wffs".

Definition 7 (The set $Cn\Gamma[PŁ4]$). *The set of consequences in PŁ4 of a set of wffs Γ (in symbols, $Cn\Gamma[PŁ4]$) is defined as follows. $Cn\Gamma[PŁ4] = \{A \mid \Gamma \vdash_{PŁ4} A\}$.*

Remark 2 ($Cn\Gamma[PŁ4]$ is a theory). *It is clear that $Cn\Gamma[PŁ4]$ is a theory.*

Lemma 1 (Extension to prime theories). *Let a be a theory and A a wff such that $A \notin a$. Then, there is a prime theory x such that $a \subseteq x$ and $A \notin x$.*

Proof. By using, for example, Kurakowski-Zorn's Lemma, we extend a to a maximal theory x such that $A \notin x$. Suppose x is not prime. Then, there are wffs B, C such that $(B \to C) \to C \in x$ but $B \notin x$ and $C \notin x$. We define the sets $[x, B] = \{D \mid B \to D \in x\}$, $[x, C] = \{D \mid C \to D \in x\}$. By using A2, it is shown that $[x, B]$ and $[x, C]$ are closed under MP; by using A1, that they include x; and finally, by T1, that $B \in [x, B]$ and $C \in [x, C]$. Next, given that $B \notin x$ and $C \notin x$, it follows that neither $[x, B]$ nor $[x, C]$ is included in x, whence by the maximality of x, $A \in [x, B]$ and $A \in [x, C]$. But then, we have $A \in x$ (by T2 and the hypothesis $(B \to C) \to C \in x$), which is impossible. Therefore, x is prime. □

Next, we build the prime theory $\mathcal{T}$ upon which the canonical model is defined.

Proposition 1 (The building of $\mathcal{T}$). *Let Γ be a set of wffs and A a wff such that $\Gamma \nvdash_{PŁ4} A$. Then, there is some prime theory $\mathcal{T}$ such that $\Gamma \subseteq \mathcal{T}$ an $A \notin \mathcal{T}$.*

Proof. For Γ and A, suppose $\Gamma \nvdash_{PŁ4} A$. Then, $A \notin Cn\Gamma[PŁ4]$. By Lemma 1, there is a prime theory $\mathcal{T}$ such that $Cn\Gamma[PŁ4] \subseteq \mathcal{T}$ and $A \notin \mathcal{T}$. (Notice that $\mathcal{T}$ is a-consistent.) □

The canonical model is defined upon the theory $\mathcal{T}$ just built as follows.

Definition 8 (The canonical 2PŁ4-model). *The canonical 2PŁ4-model is the structure $(K^C, *^C, \models^C)$, where $K^C = \{\mathcal{T}, \mathcal{T}^{*^C}\}$, $\mathcal{T}$ being the theory built up in Proposition 1, and $*^C$ and $\models^C$ be defined on K^C as follows: For any $a \in K^C$ and wff A, $a^{*^C} = \{A \mid \neg A \notin a\}$; $a \models^C A$ iff $A \in a$.*

In the sequel, it is proved that the canonical 2PŁ4-model is indeed a 2PŁ4-model. This requires to prove that $*^C$ is an involution on K^C and that $\models^C$ fulfills the conditions (i)-(iii) stated in Definition 3. (In the rest of the section, the superscript C is generally omitted above $*$ and $\models$ when there is no risk of confusion.)

Proposition 2 ($\mathcal{T}^{*^C}$ is closed under PŁ4-entailment). *The set $\mathcal{T}^{*^C}$ is closed under PŁ4-entailment, that is, if $\vdash_{PŁ4} A \to B$ and $A \in \mathcal{T}^{*^C}$, then $B \in \mathcal{T}^{*^C}$.*

Proof. Suppose (1) $\vdash_{PŁ4} A \to B$ and (2) $A \in \mathcal{T}^{*^C}$ (i.e., $\neg A \notin \mathcal{T}$). By 1 and Con_1 ($\vdash_{PŁ4} A \to B \Rightarrow \vdash_{PŁ4} \neg B \to \neg A$. Cf. the Appendix), we have (3) $\vdash_{PŁ4} \neg B \to \neg A$, whence by Corollary 1 and 2, (4) $\neg B \notin \mathcal{T}$, i.e., $B \in \mathcal{T}^{*^C}$ follows, as desired. $\square$

Proposition 3 ($*^C$ is an involutive operation on K^C). *The canonical operation $*^C$ is an involutive operation on K^C.*

Proof. It suffices to prove that $*^C$ is involutive. Let $x \in K^C$. (a) $A \in x \Rightarrow A \in x^{**}$. Let A be a wff such that $A \in x$. By A4, $\neg\neg A \in x$, whence by definition of $*^C$, firstly we get $\neg A \notin x^*$ and then $A \in x^{**}$. (b) $A \notin x \Rightarrow A \notin x^{**}$. Let A be a wff such that $A \notin x$. By A5, $\neg\neg A \notin x$, whence by definition of $*^C$, firstly we get $\neg A \in x^*$, and then $A \notin x^{**}$. $\square$

Proposition 4 (Clauses (i)-(iii) hold canonically). *Conditions (i)-(iii) in Definition 3 are satisfied by the canonical 2PŁ4-model.*

Proof. Condition (i) is trivial and Condition (iii) is immediate by Definition 8. So, let us prove Condition (ii).

(a) ($\Rightarrow$) Suppose that A and B are wffs such that $A \to B \in \mathcal{T}$ and $A \in \mathcal{T}$. Then, $B \in \mathcal{T}$ follows immediately by closure of $\mathcal{T}$ under MP. (a) ($\Leftarrow$) Suppose that A and B are wffs such that (1) $A \to B \notin \mathcal{T}$. We have to prove $A \in \mathcal{T}$ and $B \notin \mathcal{T}$. For reductio, assume (2) $A \notin \mathcal{T}$ or (3) $B \in \mathcal{T}$. By A3, $[(A \to B) \to A] \to A$, and primeness of $\mathcal{T}$, we have (4) either $A \to B \in \mathcal{T}$ or $A \in \mathcal{T}$. But 1 and 2 contradict 4. On the other hand, given 3 and A1, $B \to (A \to B)$, we get (5) $A \to B \in \mathcal{T}$, contradicting 1. Thus, $A \in \mathcal{T}$ and $B \notin \mathcal{T}$, as was to be proved.

(b) ($\Rightarrow$) Suppose that A and B are wffs such that (1) $A \to B \in \mathcal{T}^*$ (i.e., $\neg(A \to B) \notin \mathcal{T}$) and (2) $A \in \mathcal{T}^*$ (i.e., $\neg A \notin \mathcal{T}$) and, for reductio, (3) $B \notin \mathcal{T}^*$ (i.e.,

$\neg B \in \mathcal{T}$). By A8, $\neg B \to [[\neg A \to \neg(A \to B)] \to \neg(A \to B)]$ and 3, we have (4) $[\neg A \to \neg(A \to B)] \to \neg(A \to B) \in \mathcal{T}$, whence by primeness of $\mathcal{T}$, we get (5) either $\neg(A \to B) \in \mathcal{T}$ or $\neg A \in \mathcal{T}$. But 1 and 2 contradict 5. (b) ($\Leftarrow$) Suppose that A and B are wffs such that (1) $A \to B \notin \mathcal{T}^*$ (i.e., $\neg(A \to B) \in \mathcal{T}$). We have to prove $A \in \mathcal{T}^*$ (i.e., $\neg A \notin \mathcal{T}$) and $B \notin \mathcal{T}^*$ (i.e., $\neg B \in \mathcal{T}$). By A7, $\neg(A \to B) \to \neg B$ and 1, we have (2) $\neg B \in \mathcal{T}$. On the other hand, for reductio, suppose (3) $A \notin \mathcal{T}^*$ (i.e., $\neg A \in \mathcal{T}$). By A6, $\neg(A \to B) \to (\neg A \to C)$, and 3, we get $C \in \mathcal{T}$ for any wff C, contradicting the a-consistency of $\mathcal{T}$. Thus, $A \in \mathcal{T}^*$ and $B \notin \mathcal{T}^*$, as it was to be proved. □

By Propositions 3 and 4 just proved, we have the following corollary.

Corollary 2 (The canonical model is a model). *The canonical 2PŁ4-model is indeed a 2PŁ4-model.*

Proof. It follows immediately by Propositions 3 and 4. □

Finally, we prove the completeness of PŁ4 w.r.t. the 2 set-up Routley semantics.

Theorem 2 (Completeness of PŁ4). *For any set of wffs Γ and wff A, if $\Gamma \vDash_{2PL4} A$, then $\Gamma \vdash_{PL4} A$.*

Proof. Suppose $\Gamma \nvdash_{PL4} A$. By Proposition 1, there is a prime regular a-consistent theory $\mathcal{T}$ such that $\Gamma \subseteq \mathcal{T}$ and $A \notin \mathcal{T}$. Then, the canonical 2PŁ4-model is defined upon $\mathcal{T}$ as shown in Definition 8. By Corollary 2, the canonical 2PŁ4-model is a 2PŁ4-model. Then, $\Gamma \nvDash^C A$, since $\mathcal{T} \vDash^C \Gamma$ but $\mathcal{T} \nvDash^C A$. Thus, $\Gamma \nvDash_{2PL4} A$ by Definition 4. □

The section is ended by taking a look at $\mathcal{T}^*$ and its relation with $\mathcal{T}$. We begin by proving a useful proposition and a corollary thereof.

Proposition 5 (a-consistency = w-consistency). *Let a be a theory. a is a-consistent iff a is w-consistent theory (cf. Definition 6).*

Proof. ($\Rightarrow$) Suppose a is a-consistent and, for reductio, $\neg A \in a$, A being a PŁ4-theorem. Let B be an arbitrary wff. By the rule Efq$_2$, $\vdash_{PL4} A \Rightarrow \vdash_{PL4} \neg A \to B$ (cf. the Appendix), $\neg A \to B$ is a PŁ4-theorem. Then, we have $B \in a$, contradicting the a-consistency of a. ($\Leftarrow$) The proof is immediate. □

Corollary 3 ($\mathcal{T}$ is w-consistent). *The theory $\mathcal{T}$ built up in Proposition 1 is w-consistent.*

Proof. Immediate by Propositions 1 and 5. □

Proposition 6 (On the properties of $\mathcal{T}^{*^C}$). *The set $\mathcal{T}^{*^C}$ is a prime and a-consistent theory.*

Proof. (1) $\mathcal{T}^*$ is closed under MP: by using A8 similarly as in Proposition 4. (2) $\mathcal{T}^*$ contains all PŁ4-theorems: suppose A is a PŁ4-theorem such that $A \notin \mathcal{T}^*$. Then, $\neg A \in \mathcal{T}$, contradicting the w-consistency of $\mathcal{T}$. (3) $\mathcal{T}^*$ is prime: by using T3, $\neg B \to [\neg A \to \neg[(A \to B) \to B]]$ (cf. Remark 1). (4) $\mathcal{T}^*$ is a-consistent: suppose that A is a PŁ4-theorem and $\neg A \in \mathcal{T}^*$. Then, $\neg\neg A \notin \mathcal{T}$, whence by A4, $A \notin \mathcal{T}$, contradicting the fact that $\mathcal{T}$ contains all PŁ4-theorems. □

Nevertheless, we note that $\mathcal{T}$ and $\mathcal{T}^*$ are independent from each other.

Remark 3 ($\mathcal{T} \neq \mathcal{T}^*$). *The proof of $\mathcal{T} \subseteq \mathcal{T}^*$ requires consistency (in the classical sense) of $\mathcal{T}$; that of $\mathcal{T}^* \subseteq \mathcal{T}$, completeness of $\mathcal{T}$. But PŁ4 is a paraconsistent and paracomplete logic (a theory a is complete iff $A \in a$ or $\neg A \in a$ for every wff A; a is consistent in the classical sense if $A \wedge \neg A \notin a$ for every wff A).*

5 Conclusion

As pointed out in the introduction, PŁ4 is a very interesting and natural 4-valued logic. In the present paper, new light is shed on this system by endowing it with a 2 set-up Routley semantics. This semantics is fundamentally obtained by dropping the ternary relation characteristic of 2 set-up Routley-Meyer semantics as developed in [3] and [11]. Possible future work on the topic could consist in building a 2 set-up binary Routley semantics by using a binary relation on the set K, instead of a ternary one. In addition to require that this relation R be reflexive (i.e., $R00$ and $R0^*0^*$), we have essentially two classes of models: those requiring $R00^*$ and those taking $R0^*0$, since addition of both conditions would cause the collapse into classical propositional logic. In order to define 2 set-up binary Routley semantics, we could use the results in general binary Routley semantics displayed in [12].

A Appendix

The conjunction ($\wedge$), disjunction ($\vee$), necessity ($\square$) and possibility ($\diamond$) connectives given by the following tables:

$\wedge$	0	1	2	3
0	0	0	0	0
1	0	1	0	1
2	0	0	2	2
3	0	1	2	3

$\vee$	0	1	2	3
0	0	1	2	3
1	1	1	3	3
2	2	3	2	3
3	3	3	3	3

$\square$	
0	0
1	0
2	0
3	3

$\diamond$	
0	0
1	3
2	3
3	3

are definable in MPŁ4 by putting, for any wffs A, B: $A \vee B =_{df} (A \rightarrow B) \rightarrow B$; $A \wedge B =_{df} \neg(\neg A \vee \neg B)$; $\Box A =_{df} \neg(A \rightarrow \neg A)$; $\Diamond A =_{df} \neg\Box\neg A$.

Next, we remark some theorems and rules of PŁ4. Firstly, notice that any theorem of negationless classical propositional logic is a theorem of PŁ4, since the following wffs are provable in PŁ4: (t1) $A \rightarrow (A \vee B)$; (t2) $B \rightarrow (A \vee B)$; (t3) $(A \rightarrow C) \rightarrow [(B \rightarrow C) \rightarrow [(A \vee B) \rightarrow C]]$; (t4) $(A \wedge B) \rightarrow A$; (t5) $(A \wedge B) \rightarrow B$; (t6) $A \rightarrow [B \rightarrow (A \wedge B)]$. But A1-A3 (cf. §2) and t1-t6 axiomatize (together with MP) the negationless fragment of classical propositional logic. In addition, the following are also theorems and rules of PŁ4:

Con 1. $\vdash_{PŁ4} A \rightarrow B \Rightarrow \vdash_{PŁ4} \neg B \rightarrow \neg A$

Con 2. $\vdash_{PŁ4} A \rightarrow \neg B \Rightarrow \vdash_{PŁ4} B \rightarrow \neg A$

Con 3. $\vdash_{PŁ4} \neg A \rightarrow B \Rightarrow \vdash_{PŁ4} \neg B \rightarrow A$

Con 4. $\vdash_{PŁ4} \neg A \rightarrow \neg B \Rightarrow \vdash_{PŁ4} B \rightarrow A$

Efq$_1$. $\vdash_{PŁ4} \neg A \Rightarrow \vdash_{PŁ4} A \rightarrow B$

Efq$_2$. $\vdash_{PŁ4} A \Rightarrow \vdash_{PŁ4} \neg A \rightarrow B$

t7. $\neg(A \vee B) \leftrightarrow (\neg A \wedge \neg B)$

t8. $\neg(A \wedge B) \leftrightarrow (\neg A \vee \neg B)$

t9. $(A \vee B) \leftrightarrow \neg(\neg A \wedge \neg B)$

t10. $(A \wedge B) \leftrightarrow \neg(\neg A \vee \neg B)$

t11. $\Box A \leftrightarrow \neg\Diamond\neg A$

t12. $\Diamond A \leftrightarrow \neg\Box\neg A$

t13. $\Box A \rightarrow A$

t14. $A \rightarrow \Diamond A$

t15. $\Box A \rightarrow \Box\Box A$

t16. $\Diamond A \rightarrow \Box\Diamond A$

t17. $\Diamond\Box A \rightarrow \Box A$

t18. $\Box(A \rightarrow B) \rightarrow (\Box A \rightarrow \Box B)$

t19. $\Box(A \wedge B) \leftrightarrow (\Box A \wedge \Box B)$

t20. $\Diamond(A \vee B) \leftrightarrow (\Diamond A \vee \Diamond B)$

t21. $\Diamond(A \rightarrow B) \leftrightarrow (\Box A \rightarrow \Diamond B)$

t22. $(\Diamond A \rightarrow \Box B) \rightarrow \Box(A \rightarrow B)$

t23. $(\Diamond A \rightarrow \Diamond B) \rightarrow \Diamond(A \rightarrow B)$

t24. $(\Box A \vee \Box B) \rightarrow \Box(A \vee B)$

t25. $(\Diamond A \wedge \Diamond B) \to \Diamond(A \wedge B)$

t26. $\Box(A \vee B) \to (\Box A \vee \Diamond B)$

t27. $(\Diamond A \wedge \Box B) \to \Diamond(A \wedge B)$

t28. $A \vee \neg \Box A$

t29. $(\Box A \wedge \neg A) \to B$

t30. $A \to (\neg A \vee \Box A)$

Nec. $\vdash_{PŁ4} A \Rightarrow \vdash_{PŁ4} \Box A$

RT. $\vdash_{PŁ4} A \leftrightarrow B \Rightarrow \vdash_{PŁ4} C[A] \leftrightarrow C[A/B]$

DT. $\Gamma, A \vdash_{PŁ4} B \Rightarrow \Gamma \vdash_{PŁ4} A \to B$

(The biconditional ($\leftrightarrow$) is defined in the customary way: $A \leftrightarrow B =_{df} (A \to B) \wedge (B \to A)$. Con abbreviates Contraposition. Efq abbreviates 'E falso quodlibet' — Any proposition is implied by a false proposition. Nec abbreviates 'Necessitation' rule. RT abbreviates 'Replacement theorem': $C[A]$ is a wff where A appears; $C[A/B]$ is the result of changing one or more occurrences of A in $C[A]$ for corresponding occurrences of B. Finally, DT means 'Deduction Theorem'.)

References

[1] Anderson, A. R., Belnap, N. D. Jr., Dunn, J. M. (1992). *Entailment, The Logic of Relevance and Necessity*, Vol. II. Princeton University Press.

[2] Béziau, J.-Y. (2011). A new four-valued approach to modal logic. *Logique et Analyse*, 54(213), 109-121.

[3] Brady, R. T. (1982). Completeness proofs for the systems RM3 and BN4. *Logique et Analyse*, 25, 9-32.

[4] De, M., Omori, H. (2015). Classical negation and expansions of Belnap–Dunn logic. *Studia Logica*, 103(4), 825-851. https://doi.org/10.1007/s11225-014-9595-7.

[5] Kamide, N., Omori, H. (2017). An extended first-order Belnap-Dunn logic with classical negation. In A. Baltag, J. Seligman, & T. Yamada (Eds.), *Logic, Rationality, and Interaction* (pp. 79-93). Springer. https://doi.org/10.1007/978-3-662-55665-8_6.

[6] Łukasiewicz, J. (1951). *Aristotle's syllogistic from the standpoint of modern formal logic*. Clarendon Press, Oxford.

[7] Łukasiewicz, J. (1953). A system of modal logic. *The Journal of Computing Systems*, 1, 111-149.

[8] Méndez, J. M., Robles, G. (2015). A strong and rich 4-valued modal logic without Łukasiewicz-Type Paradoxes. *Logica Universalis*, 9(4), 501-522. https://doi.org/10.1007/s11787-015-0130-z.

[9] Méndez, J. M., Robles, G., Salto, F. (2016). An interpretation of Łukasiewicz's 4-valued modal logic. *Journal of Philosophical Logic*, 45(1). 73-87. https://doi.org/10.1007/s10992-015-9362-x.

[10] Omori, H., Skurt, D. (2019). SIXTEEN$_3$ in light of Routley stars. In R. Iemhoff, M. Moortgat, & R. de Queiroz (Eds.), *Logic, Language, Information, and Computation, WoLLIC 2019* (pp. 516-532). Springer. https://doi.org/10.1007/978-3-662-59533-6_31

[11] Robles, G., López, S. M., Blanco, J. M., Recio, M. M., Paradela, J. R. (2016). A 2-set-up Routley-Meyer semantics for the 4-valued relevant logic E4. *Bulletin of the Section of Logic*, 45(2), 93-109. https://doi.org/10.18778/0138-0680.45.2.03.

[12] Robles, G., Méndez, J. M. (2015). A binary Routley semantics for intuitionistic de morgan minimal logic H$_M$ and its extensions. *Logic Journal of the IGPL*, 23(2), 174-193. https://doi.org/10.1093/jigpal/jzu029.

[13] Robles, G., Méndez, J. M. (2018). *Routley-Meyer ternary relational semantics for intuitionistic-type negations.* Elsevier. https://doi.org/10.1016/C2015-0-01638-0

[14] Routley, R., Routley, V. (1972). The semantics of First Degree Entailment. *Noûs*, 6(4), 335-359. https://doi.org/10.2307/2214309.

[15] Routley, R., Meyer, R. K. (1973). The semantics of entailment I. In H. Leblanc (Ed.), *Truth, syntax and modality. Proceedings of the Temple University Conference on Alternative Semantics. Studies in logic and the foundations of mathematics*, vol. 68. (pp. 199–243). North-Holland Publishing Company, Amsterdam and London.

[16] Routley, R. Meyer, R. K., Plumwood, V., Brady R. T. (1982). *Relevant logics and their rivals*, vol. 1. Atascadero, CA: Ridgeview Publishing Co.

[17] Zaitsev, D. (2012). *Generalized relevant logic and models of reasoning.* [Doctoral dissertation]. Moscow State Lomonosov University.

The Embedding Path Order for Lambda-Free Higher-Order Terms

Alexander Bentkamp
Vrije Universiteit Amsterdam, Netherlands
State Key Lab. of Computer Science, Institute of Software, CAS, Beijing, China
bentkamp@gmail.com

Abstract

The embedding path order, introduced in this article, is a variant of the recursive path order (RPO) for untyped λ-free higher-order terms (also called applicative first-order terms). Unlike other higher-order variants of RPO, it is a ground-total and well-founded simplification order, making it more suitable for the superposition calculus. I formally proved the order's theoretical properties in Isabelle/HOL and evaluated the order in a prototype based on the superposition prover Zipperposition.

1 Introduction

Superposition [2] is one of the most successful calculi for proof search in first-order logic with equality. To restrict the search space, it uses a term order, which in practice is usually the Knuth–Bendix order (KBO) [24] or the recursive path order (RPO) [16]. Although, in isolation, KBO often achieves better results, modern portfolio provers employ both KBO and RPO in separate proof attempts because the two orders complement each other well.

With colleagues, I have developed a superposition-like calculus for λ-free higher-order logic (also called applicative first-order logic) [6]. Supporting partial applications and applied variables, this logic allows for terms such as f a b, f a, f, and x a b. To apply superposition to this logic, the term order must be generalized. For KBO, a suitable generalization is λ-free higher-order KBO (λfKBO) [3]. It is a ground-total and well-founded simplification order, and therefore a straight-forward generalization of superposition is refutationally complete—i.e., the generalized calculus will eventually find a proof for any given theorem. This approach has been implemented in the E prover [36].

In contrast, RPO's generalization to λ-free higher-order terms, λfRPO [9], is nonmonotonic—i.e. it lacks the property that $t > s$ implies $u[t] > u[s]$. Without monotonicity, the straight-forward generalization of superposition is not refutationally complete. In our work on superposition for λ-free higher-order logic, we have shown that performing additional inferences can recover refutational completeness for nonmonotonic orders. Our calculus for nonmonotonic orders has promising evaluation results, but the theory and implementation of the calculus is rather complex. We asked ourselves:

> *Is there an RPO-like ground-total and well-founded simplification order for lambda-free higher-order terms?*

If "RPO-like" means that the order must coincide with RPO on the first-order fragment of lambda-free higher-order logic, then the answer is no, as the following example shows: If $g \succ f \succ b \succ a$, then $g\ b > f\ (g\ a)\ b$ by coincidence with first-order RPO, corresponding to $g(b) > f(g(a), b)$ in first-order syntax, but $g < f\ (g\ a)$ by the subterm property and hence $g\ b < f\ (g\ a)\ b$ by monotonicity, yielding a contradiction.

If "RPO-like" means that the order should merely resemble RPO, the answer is yes. One candidate is the applicative RPO, which is obtained by encoding λ-free higher-order terms applicatively into first-order logic via a binary symbol app representing application—e.g. $x\ a\ b$ as $app(app(x, a), b)$—and then using first-order RPO. However, with this approach the symbol app becomes pervasive, which undermines RPO's principle of comparing the precedence of different symbols. Moreover, it is impossible to assign different extension orders such as the lexicographic or multiset extension to different function symbols because the only applied function symbol in the encoding is app.

This article presents an answer to our question that avoids the applicative encoding: the embedding path order (EPO[1]). It supports different extension operators for different function symbols (Section 3). The main difference to RPO lies in using the notion of embeddings where RPO uses the notion of direct subterms (Section 4). EPO is a ground-total and well-founded simplification order and I have formally proved this property in Isabelle/HOL (Section 5). Thus EPO allows us to avoid the theoretical and implementational challenges that λfRPO poses.

However, a good term order for superposition must also be efficient to compute. I have implemented EPO as a prototype in the superposition prover Zipperposition (Section 6). Its worst-case time complexity is quintic, and thus slower than for KBO and RPO, which can be computed in linear and quadratic time, respectively. I

---

[1]Beware that the unrelated exptime path order [19] has the same abbreviation.

evaluate the use of EPO for superposition on TPTP [35] and Sledgehammer [32] benchmarks and compare it with λfRPO, using our approach for nonmonotonic orders (Section 7). The results show that the approach with the nonmonotonic λfRPO performs slightly better. Nonetheless, EPO offers a way to complement λfKBO with much less implementation effort in provers that currently support only monotonic orders, such as the E prover.

An earlier version of this article is part of my PhD thesis [5].

2 Preliminaries

We fix a set of variables $\mathcal{V}$ and a nonempty (possibly infinite) set of symbols Σ. We reserve the names x, y, z for variables and a, b, c, f, g, h for symbols.

In untyped λ-free higher-order logic, a term is defined inductively as being either a variable, a symbol, or an application $s\,t$, where s and t are terms.

These terms are isomorphic to applicative terms [23]. We reserve the names t, s, v, u for terms and use $\mathcal{T}$ to denote the set of all terms. Application is left-associative, i.e., $s\,t\,u = (s\,t)\,u$. Any term can be written as $\zeta\,\bar{t}_n$ using spine notation [13], where ζ is a nonapplication term, called *head*, and $\bar{t}_n$ is a tuple of terms, called *arguments*. It represents the term $\zeta\,t_1\,\ldots\,t_n$. Here and elsewhere, $\bar{t}_n$ or $\bar{t}$ stands for the tuple $(t_1, \ldots, t_n)$. We write $()$ for the empty tuple, t for the singleton tuple (t), and $\bar{s} \cdot \bar{t}$ for the concatenation of the tuples $\bar{s}$ and $\bar{t}$.

The *size* $|t|$ of a term t is inductively defined as 1 if $t \in \mathcal{V} \cup \Sigma$ and as $|t_1| + |t_2|$ if t is an application $t_1\,t_2$. A *subterm* of a term t is inductively defined as being either t itself or, if t is an application $t_1\,t_2$, a subterm of t_1 or of t_2.

The embedding relation [1, Definition 5.4.2] is a generalization of the subterm relation: First, the embedding step relation $\rightarrow_{\mathsf{emb}}$ is inductively defined as follows. For any terms s, t, and t', we have $t\,s \rightarrow_{\mathsf{emb}} t$ and $t\,s \rightarrow_{\mathsf{emb}} s$; and if $t \rightarrow_{\mathsf{emb}} t'$, then $t\,s \rightarrow_{\mathsf{emb}} t'\,s$ and $s\,t \rightarrow_{\mathsf{emb}} s\,t'$. For example, f a b c d $\rightarrow_{\mathsf{emb}}$ a b c d, f a b c d $\rightarrow_{\mathsf{emb}}$ f a c d, and f (g (h a) b) c $\rightarrow_{\mathsf{emb}}$ f (g h b) c. Let the embedding relation $\trianglerighteq_{\mathsf{emb}}$ be the reflexive transitive closure of $\rightarrow_{\mathsf{emb}}$.

Given a binary relation $>$, we write $<$ for its converse (i.e., $a < b \Leftrightarrow b > a$) and $\geq$ for its reflexive closure (i.e., $b \geq a \Leftrightarrow b > a \vee b = a$). A binary relation $>$ on λ-free higher-order terms is a *simplification order* if it is irreflexive (i.e., $t \not> t$), is transitive (i.e., $u > t > s \Rightarrow u > s$), is monotonic (i.e., $t > s \Rightarrow u\,t > u\,s \wedge t\,u > s\,u$), is stable under substitutions (i.e., $t > s \Rightarrow t\sigma > s\sigma$), and has the subterm property (i.e., $t \geq s$ if s is a subterm of t). It is *ground-total* if for all distinct ground terms s and t either $t > s$ or $t < s$. It is *well founded* if there is no infinite descending chain $t_1 > t_2 > \cdots$.

We view RPO as a term order on the first-order fragment of λ-free higher-order terms, identifying first-order terms $\mathsf{f}(\bar{t})$ with $\mathsf{f}\,\bar{t}$. Let $\succ$ be a well-founded total order on Σ. Then RPO is inductively defined as follows: $t >_{\mathsf{rp}} s$ if any of the following conditions are met, where $t = \mathsf{g}\,\bar{t}$ and $s = \mathsf{f}\,\bar{s}$:

R1. $s \in \mathcal{V}$, $t \neq s$, and s occurs in t;

R2. $t_i \geq_{\mathsf{rp}} s$ for some i;

R3. $\mathsf{g} \succ \mathsf{f}$ and $t >_{\mathsf{rp}} s_i$ for all i;

R4. $\mathsf{g} = \mathsf{f}$, $\bar{t}_n \gg^{\mathsf{f}}_{\mathsf{rp}} \bar{s}_m$, and $t >_{\mathsf{rp}} s_i$ for all i.

where is $\gg^{\mathsf{f}}_{\mathsf{rp}}$ is an extension of $>_{\mathsf{rp}}$ to tuples—e.g., the lexicographic extension or the multiset extension. I will present a more formal definition of extension operators $> \mapsto \gg$ in the following section.

3 Extension operators

In the spirit of RPO, EPO compares the heads of terms and, in case of equality, proceeds to compare the argument tuples. There is a variety of ways to extend a binary relation $>$ on an arbitrary set A to a binary relation $\gg$ on tuples A^*, which we call extension operators. We define extension operators on binary relations, not on partial orders, because they are used in the definition of EPO at a point where we have not shown EPO to be a partial order yet.

Definition 1. We define the following properties of extension operators $> \mapsto \gg$, which are required for EPO to be a ground-total and well-founded simplification order. Here, given a function $h : A \to A$, we write $h(\bar{a})$ for the componentwise application of h to $\bar{a}$.

X1. Monotonicity:
$\bar{b} \gg_1 \bar{a}$ implies $\bar{b} \gg_2 \bar{a}$ if for all $a, b \in A$, $b >_1 a$ implies $b >_2 a$

X2. Preservation of stability:
$\bar{b} \gg \bar{a}$ implies $h(\bar{b}) \gg h(\bar{a})$ if for all $a, b \in \bar{a} \cup \bar{b}$, $b > a$ implies $h(b) > h(a)$

X3. Preservation of transitivity: $\gg$ is transitive if $>$ is transitive

X4. Preservation of irreflexivity:
$\gg$ is irreflexive if $>$ is irreflexive and transitive

X5. Preservation of well-foundedness: $\gg$ is well founded if $>$ is well founded

X6. Compatibility with tuple contexts: $b > a$ implies $\bar{c} \cdot b \cdot \bar{d} \gg \bar{c} \cdot a \cdot \bar{d}$

X7. Preservation of totality: $\gg$ is total if $>$ is total

X8. Compatibility with prepending: $\bar{b} \gg \bar{a}$ implies $c \cdot \bar{b} \gg c \cdot \bar{a}$

X9. Compatibility with appending: $\bar{b} \gg \bar{a}$ implies $\bar{b} \cdot c \gg \bar{a} \cdot c$

X10. Minimality of the empty tuple: $a \gg ()$ for all $a \in A$

The length-lexicographic extension operator, left-to-right or right-to-left, fulfills all these properties:

Definition 2. The *left-to-right length-lexicographic extension operator* $> \mapsto \gg^{\text{ltr}}$ is defined inductively as follows: $\bar{a}_m \gg^{\text{ltr}} \bar{b}_n$ if $m > n$; or $m = n > 0$ and $a_1 > b_1$; or $m = n > 0$, $a_1 = b_1$, and $(a_2, \ldots, a_m) \gg^{\text{ltr}} (b_2, \ldots, b_n)$. The *right-to-left length-lexicographic extension operator* $> \mapsto \gg^{\text{rtl}}$ is defined inductively as follows: $\bar{a}_m \gg^{\text{rtl}} \bar{b}_n$ if $m > n$; or $m = n > 0$ and $a_m > b_n$; or $m = n > 0$, $a_m = b_n$, and $(a_1, \ldots, a_{m-1}) \gg^{\text{rtl}} (b_1, \ldots, b_{n-1})$.

The multiset extension operator fulfills all properties except X7, but if combined with a lexicographic comparison as a tie-breaker, it fulfills all properties as well:

Definition 3. The *multiset extension operator with tie-breaker* $> \mapsto \gg^{\text{ms}}$ is defined as follows: $\bar{a} \gg^{\text{ms}} \bar{b}$ if the multiset containing the elements of $\bar{a}$ is larger than the multiset containing the elements of $\bar{b}$ with respect to Dershowitz and Manna's multiset order [18]; or if the two multisets are equal and $\bar{a} \gg^{\text{ltr}} \bar{b}$.

Blanchette et al. [9] give a more detailed account of different extension operators. Their list of properties is identical with the one above, except for X2, which they originally formulated differently but corrected in their technical report [8].

4 The order

Any simplification order has the embedding property, i.e., the property that $t \trianglerighteq_{\text{emb}} s$ implies $t \succeq s$ [1, Lemma 5.4.7]. The fundamental idea of EPO is to enforce the embedding property by replacing the notion of subterms used in the definition of RPO by the notion of embeddings. Performed naively, this causes issues with stability under substitution and with the time complexity of the order computation due to the large number of possible embedding steps. Both of these issues are addressed by EPO.

Definition 4 (EPO). Let $\succ$ be a well-founded total order on Σ. For each $f \in \Sigma$, let $> \mapsto \gg^f$ be an extension operator satisfying the properties of Definition 1. The induced *embedding path order* $>_{\text{ep}}$ is inductively defined as follows: $t >_{\text{ep}} s$ if any of the following conditions is met, where $t = \xi\, \bar{t}_n$ and $s = \zeta\, \bar{s}_m$:

E1. $n > 0$ and $\mathit{chop}(t) \geq_{\text{ep}} s$

E2. $\xi, \zeta \in \Sigma$, $\xi \succ \zeta$, and either $m = 0$ or $t >_{\text{ep}} \mathit{chop}(s)$

E3. $\xi, \zeta \in \Sigma$, $\xi = \zeta$, $\bar{t}_n \gg_{\text{ep}}^{\zeta} \bar{s}_m$, and either $m = 0$ or $t >_{\text{ep}} \mathit{chop}(s)$

E4. $\xi, \zeta \in \mathcal{V}$, $\xi = \zeta$, $\bar{t}_n \gg_{\text{ep}}^f \bar{s}_m$ for all $f \in \Sigma$, $n > 0$, and either $m = 0$ or $\mathit{chop}(t) >_{\text{ep}} \mathit{chop}(s)$

Here, for a term $\xi\, \bar{t}_n$ with $n > 0$, we define $\mathit{chop}(\xi\, \bar{t}_n)$ as the term resulting from applying t_1 to the remaining arguments, i.e., $\mathit{chop}(\xi\, \bar{t}_n) = t_1\, t_2\, \ldots\, t_n$. (For example, $\mathit{chop}(\mathsf{f}\, (\mathsf{g}\, \mathsf{a})\, (\mathsf{h}\, \mathsf{b})) = \mathsf{g}\, \mathsf{a}\, (\mathsf{h}\, \mathsf{b})$.)

The following examples illustrate the differences between RPO and EPO on first-order terms. We use the precedence $\mathsf{g} \succ \mathsf{f} \succ \mathsf{c} \succ \mathsf{b} \succ \mathsf{a}$ and the left-to-right length-lexicographic extension for both orders.

$$\mathsf{f}\,(\mathsf{g}\,\mathsf{a})\,\mathsf{b} <_{\text{rp}} \mathsf{g}\,\mathsf{b} \qquad \mathsf{f}\,(\mathsf{g}\,\mathsf{a})\,\mathsf{c} <_{\text{rp}} \mathsf{g}\,\mathsf{b} \qquad \mathsf{g}\,x\,y >_{\text{rp}} \mathsf{f}\,y\,y$$
$$\mathsf{f}\,(\mathsf{g}\,\mathsf{a})\,\mathsf{b} >_{\text{ep}} \mathsf{g}\,\mathsf{b} \qquad \mathsf{f}\,(\mathsf{g}\,\mathsf{a})\,\mathsf{c} >_{\text{ep}} \mathsf{g}\,\mathsf{b} \qquad \mathsf{g}\,x\,y \not>_{\text{ep}} \mathsf{f}\,y\,y$$

The first term pair illustrates that RPO does not have the embedding property, whereas EPO does. The relation $\mathsf{f}\,(\mathsf{g}\,\mathsf{a})\,\mathsf{b} >_{\text{ep}} \mathsf{g}\,\mathsf{b}$ can be shown by applying E1. E1 requires $\mathsf{g}\,\mathsf{a}\,\mathsf{b} >_{\text{ep}} \mathsf{g}\,\mathsf{b}$, which holds by E3. Finally we need E2 to show $\mathsf{g}\,\mathsf{a}\,\mathsf{b} >_{\text{ep}} \mathsf{b}$. The second term pair shows that there are further disagreements between the two orders, even if one term is not an embedding of the other. As above, $\mathsf{f}\,(\mathsf{g}\,\mathsf{a})\,\mathsf{c} >_{\text{ep}} \mathsf{g}\,\mathsf{b}$ can be established by applying E1, followed by E3 and E2. The third term pair is comparable with RPO but incomparable with EPO. In general, EPO cannot judge a term to be smaller if it contains more occurrences of a variable. I conjecture that there are no first-order terms comparable with EPO but incomparable with RPO. In this sense, EPO is weaker than RPO on first-order terms.

4.1 Rationale of the Definition

The definition of EPO has been carefully designed to make EPO a ground-total and well-founded simplification order that can be computed in polynomial time with respect to the size of the compared terms.

Condition E1 enforces the embedding property in a similar way as RPO's condition R2 enforces the subterm property. This underlying idea gives EPO its name. A naive approach would be to test all embedding steps to enforce the embedding property, but it is sufficient to test only the embedding step *chop*, yielding a better computational complexity. The remaining conditions follow a similar structure as RPO, but contain subconditions on *chop* where RPO has subconditions on subterms.

To achieve stability under substitutions, it is essential to demand $chop(t) >_{ep} chop(s)$ instead of $t >_{ep} chop(s)$ in E4, as the following examples show. If $>'_{ep}$ is the relation obtained from $>_{ep}$ by replacing '$chop(t)$' by 't' in E4, then we have

$$x \mathsf{f} \mathsf{f} >'_{ep} x\, x, \text{ but } \mathsf{f}\, y\, \mathsf{f} \mathsf{f} \not>'_{ep} \mathsf{f}\, y\, (\mathsf{f}\, y) \qquad x\, \mathsf{f}\, x >'_{ep} x\, (x\, \mathsf{f}), \text{ but } y\, \mathsf{f} \mathsf{f}\, (y\, \mathsf{f}) \not>'_{ep} y\, \mathsf{f}\, (y\, \mathsf{f} \mathsf{f})$$

Using $>_{ep}$, all of these pairs are incomparable.

In condition E4, it is crucial to check $\bar{t}_n \gg^{\mathsf{f}}_{ep} \bar{s}_m$ for all $\mathsf{f} \in \Sigma$. In contrast, λfKBO [3] and λfRPO [9] allow us to use a map *ghd* from variables to possible ground heads that might occur when a variable is instantiated. The corresponding condition in these orders then states '$\bar{t}_n \gg^{\mathsf{f}}_{ep} \bar{s}_m$ for all $\mathsf{f} \in ghd(\zeta)$'. For EPO, this approach cannot be used. For example, assume $\mathsf{b} \succ \mathsf{a}$, $ghd(x) = \{\mathsf{f}\}$, and that f uses the left-to-right length-lexicographic extension. Then we would have $x\, \mathsf{b}\, \mathsf{a} > x\, \mathsf{a}\, \mathsf{b}$ if we checked only the extension orders for $ghd(x)$. This contradicts stability under substitutions because, if g uses the right-to-left length-lexicographic extension, $y\, \mathsf{g}\, \mathsf{b}\, \mathsf{a}$ and $y\, \mathsf{g}\, \mathsf{a}\, \mathsf{b}$ are incomparable, assuming $ghd(y) = \{\mathsf{f}\}$.

EPO is not a simplification order when (nonlength-)lexicographic extensions are used. With the left-to-right lexicographic extension, it is nonmonotonic because for $\mathsf{g} \succ \mathsf{f} \succ \mathsf{b} \succ \mathsf{a}$, we have $\mathsf{f}\, (\mathsf{g}\, \mathsf{a}) >_{ep} \mathsf{g}$ but $\mathsf{f}\, (\mathsf{g}\, \mathsf{a})\, \mathsf{b} <_{ep} \mathsf{g}\, \mathsf{b}$. With the right-to-left lexicographic extension, it lacks stability under substitutions because $x\, \mathsf{f} > x$ but $\mathsf{f}\, y\, \mathsf{f} \not> \mathsf{f}\, y$. With the right-to-left lexicographic extension, it also lacks well-foundedness because for $\mathsf{f} \succ \mathsf{b} \succ \mathsf{a}$, we have $\mathsf{f}\, \mathsf{b} >_{ep} \mathsf{f}\, \mathsf{b}\, \mathsf{a} >_{ep} \mathsf{f}\, \mathsf{b}\, \mathsf{a}\, \mathsf{a} >_{ep} \cdots$.

4.2 In-Depth Example

The following example illustrates the benefits of EPO for superposition. Consider the following term rewriting system:

$$\mathsf{f}\, x\, \mathsf{Nil} \xrightarrow{1} x \qquad \mathsf{f}\, x\, (\mathsf{A}\, y) \xrightarrow{2} \mathsf{f}\, (\mathsf{A}\, (\mathsf{B}\, x))\, y \qquad \mathsf{f}\, x\, (\mathsf{B}\, y) \xrightarrow{3} \mathsf{f}\, (\mathsf{B}\, (\mathsf{A}\, x))\, y$$

This rewriting system can be interpreted as a definition of a function on strings. In this interpretation, Nil represents the empty string, and chains of applications of the functions A and B to Nil represent strings over the alphabet $\{\mathsf{A}, \mathsf{B}\}$; thus, A (B (B Nil)) represents the string ABB. The function f takes two such strings, reverses the second string, replaces in the resulting string each A by AB and each B by BA, and finally appends the first string.

All three rules are orientable by EPO with the right-to-left length-lexicographic extension for f and precedence $f \succ A, B$. To show that rule 1 can be oriented, we apply E1. To do so, we need to prove x Nil $>_{ep} x$, which holds by E4. To show that rule 2 can be oriented, we apply E3. To do so, we need to prove $(x, A\ y) \gg^f_{ep} ((A\ (B\ x)), y)$ and $f\ x\ (A\ y) >_{ep} A\ (B\ x)\ y$. The former holds by the definition of the right-to-left length-lexicographic extension and by E1. For the latter, we apply E2. To show $f\ x\ (A\ y) >_{ep} B\ x\ y$, we apply E2 again. To show $f\ x\ (A\ y) >_{ep} x\ y$, we apply E1. To show $x\ (A\ y) >_{ep} x\ y$, we apply E4. Finally, $A\ y >_{ep} y$ holds by E1. The proof for rule 3 is analogous.

To my knowledge, the literature contains no other ground-total simplification order for λ-free higher-order terms that can orient all three of these rules. Rules 2 and 3 are not orientable by applicative KBO or applicative RPO. With applicative KBO, the weight of the right-hand sides is always too large. With applicative RPO, too many heads are the application symbol app, preventing us from finding an appropriate precedence. With λfKBO [3], one of the two rules 2 and 3 can be oriented by assigning either A or B zero weight, but the system as a whole is not orientable with this order either. With λfRPO [9], we can orient all three rules, but λfRPO is not a simplification order.

This rewriting system suggests that EPO with a right-to-left length-lexicographic extension is generally stronger than left-to-right. If the two arguments of f were swapped, one would intuitively attempt to use the left-to-right extension for f, but fail because $f\ (A\ y)\ x \not>_{ep} y\ (A\ (B\ x))$. For this system with the arguments of f swapped, applicative RPO can orient all three rules. However, swapping arguments cannot be used as a general approach to orient rewriting systems if the affected function appears unapplied.

The term order's ability to orient equations in the right way can have considerable effects on the performance of superposition provers. Consider the rewrite rules above, recast as equations, and the negated conjecture given below, for some $k \in \mathbb{N}$:

$$f\ x\ \text{Nil} \approx x \qquad f\ x\ (A\ y) \approx f\ (A\ (B\ x))\ y \qquad f\ x\ (B\ y) \approx f\ (B\ (A\ x))\ y$$

$$f\ c\ (AB)^{k+1} \not\approx B\ (A\ (f\ c\ ((AB)^k A)))$$

where the abbreviation $(AB)^{k+1}$ stands for $A\ (B\ \ldots (A\ (B\ \text{Nil}))\ldots)$ and $(AB)^k A$ for $A\ (B\ \ldots (A\ \text{Nil})\ldots)$. Using the EPO above that can orient the equations left to right, superposition provers can solve this problem by simplification rules only. Simplification rules are much more efficient than inference rules because simplifications replace clauses and do not add new ones. Using an order that can orient only the first equation from left to right, we would need at least k inferences; using an order that can orient the first equation and only one of the other two, we would need at least $k/2$ inferences.

5 Properties of the order

EPO fulfills all the properties of a ground-total and well-founded simplification order. The proofs in this section have been developed in Isabelle/HOL and published in the Archive of Formal Proofs [4]. They are inspired by the corresponding proofs about λfRPO [9], which in turn were adapted from Baader and Nipkow [1] and Zantema [37].

Theorem 5 (Transitivity). $u >_{\text{ep}} t$ and $t >_{\text{ep}} s$ implies $u >_{\text{ep}} s$.

Proof. By well-founded induction on the multiset $\{|u|, |t|, |s|\}$ with respect to the multiset extension [18] of $>$ on $\mathbb{N}$. Let $u = \psi\,\bar{u}_r$, $t = \xi\,\bar{t}_n$ and $s = \zeta\,\bar{s}_m$.

If $u >_{\text{ep}} t$ is derived by E1, then $r > 0$ and $\mathit{chop}(u) \geq_{\text{ep}} t$. Applying the induction hypothesis to $\mathit{chop}(u)$, t, s, it follows that $\mathit{chop}(u) >_{\text{ep}} s$ and hence $u >_{\text{ep}} s$ by E1.

If $u >_{\text{ep}} t$ is derived by E2 or E3 and $t >_{\text{ep}} s$ is derived by E1, then $n > 0$ and $u >_{\text{ep}} \mathit{chop}(t) \geq_{\text{ep}} s$. Applying the induction hypothesis to u, $\mathit{chop}(t)$, s, it follows that $u >_{\text{ep}} s$.

If $u >_{\text{ep}} t$ is derived by E4 and $t >_{\text{ep}} s$ is derived by E1, then $r > 0$, $n > 0$, and $\mathit{chop}(u) >_{\text{ep}} \mathit{chop}(t) \geq_{\text{ep}} s$. By applying the induction hypothesis to $\mathit{chop}(u)$, $\mathit{chop}(t)$, s, we get $\mathit{chop}(u) >_{\text{ep}} s$. By E1, it follows that $u >_{\text{ep}} s$.

If $u >_{\text{ep}} t$ and $t >_{\text{ep}} s$ are derived by E2 and E2, by E2 and E3, or by E3 and E2, respectively, then $\psi \succ \zeta$ and $t >_{\text{ep}} \mathit{chop}(s)$. If $m = 0$, we can apply E2 directly to obtain $u >_{\text{ep}} s$. If $m > 0$, by the induction hypothesis for u, t, $\mathit{chop}(s)$, it follows from $u >_{\text{ep}} t$ and $t >_{\text{ep}} \mathit{chop}(s)$ that $u >_{\text{ep}} \mathit{chop}(s)$. Then we can apply E2 to obtain $u >_{\text{ep}} s$.

If $u >_{\text{ep}} t$ and $t >_{\text{ep}} s$ are both derived by E3, then $\psi = \xi = \zeta \in \Sigma$, $\bar{u} \gg_{\text{ep}}^{\xi} \bar{t}$, $\bar{t} \gg_{\text{ep}}^{\zeta} \bar{s}$, and either $m = 0$ or $t >_{\text{ep}} \mathit{chop}(s)$. By the induction hypothesis and by preservation of transitivity (property X3) on the set consisting of the elements of $\bar{u}$, $\bar{t}$ and $\bar{s}$, it follows that $\bar{u} \gg_{\text{ep}}^{\zeta} \bar{s}$. If $m = 0$, we obtain $u >_{\text{ep}} s$ directly by E3. If $m > 0$, we have $t >_{\text{ep}} \mathit{chop}(s)$ and by applying the induction hypothesis to u, t, $\mathit{chop}(s)$, it follows that $u >_{\text{ep}} \mathit{chop}(s)$. By E3, we have $u >_{\text{ep}} s$.

If $u >_{\text{ep}} t$ and $t >_{\text{ep}} s$ are both derived by E4, then $\psi = \xi = \zeta \in \Sigma$, $\bar{u} \gg_{\text{ep}}^{\mathsf{f}} \bar{t}$, $\bar{t} \gg_{\text{ep}}^{\mathsf{f}} \bar{s}$ for all $\mathsf{f} \in \Sigma$, $r > 0$, $n > 0$, $\mathit{chop}(u) >_{\text{ep}} \mathit{chop}(t)$, and either $m = 0$ or $\mathit{chop}(t) >_{\text{ep}} \mathit{chop}(s)$. As above, by the induction hypothesis and by preservation of transitivity (property X3) on the set consisting of the elements of $\bar{u}$, $\bar{t}$ and $\bar{s}$, it follows that $\bar{u} \gg_{\mathsf{ep}}^{\mathsf{f}} \bar{s}$ for all $\mathsf{f} \in \Sigma$. If $m = 0$, we obtain $u >_{\mathsf{ep}} s$ directly by E4. If $m > 0$, we have $\mathit{chop}(u) >_{\mathsf{ep}} \mathit{chop}(t) >_{\mathsf{ep}} \mathit{chop}(s)$. By applying the induction hypothesis to $\mathit{chop}(u)$, $\mathit{chop}(t)$, $\mathit{chop}(s)$, it follows that $\mathit{chop}(u) >_{\mathsf{ep}} \mathit{chop}(s)$. By E4, we have $u >_{\mathsf{ep}} s$.

If one of the inequalities $u >_{\mathsf{ep}} t$ and $t >_{\mathsf{ep}} s$ is derived by E2 or E3, the other cannot be derived by E4 because ξ must be either a variable or a symbol. □

Theorem 6 (Irreflexivity). $s \not>_{\mathsf{ep}} s$.

Proof. By strong induction on $|s|$. We suppose that $s >_{\mathsf{ep}} s$ and derive a contradiction. Let $s = \zeta \, \bar{s}_m$.

If $s >_{\mathsf{ep}} s$ is derived by E1, then $m > 0$ and $\mathit{chop}(s) \geq_{\mathsf{ep}} s$. From the definition of chop, it is clear that $\mathit{chop}(s) \neq s$. Hence, $\mathit{chop}(s) >_{\mathsf{ep}} s$. By E1, we have $s >_{\mathsf{ep}} \mathit{chop}(s)$. By transitivity (Theorem 5), it follows that $\mathit{chop}(s) >_{\mathsf{ep}} \mathit{chop}(s)$, which contradicts the induction hypothesis.

If $s >_{\mathsf{ep}} s$ is derived by E2, we have $\zeta \succ \zeta$, in contradiction to $\succ$ being a total order.

If $s >_{\mathsf{ep}} s$ is derived by E3 or E4, we have $\bar{s} \gg_{\mathsf{ep}}^{\mathsf{f}} \bar{s}$ for some $\mathsf{f} \in \Sigma$. By preservation of irreflexivity (property X4) on the set consisting of the elements of $\bar{s}$ and by transitivity of $>_{\mathsf{ep}}$ (Theorem 5), it follows that $s' >_{\mathsf{ep}} s'$ for some $s' \in \bar{s}$. This contradicts the induction hypothesis. □

Lemma 7. $t\, u >_{\mathsf{ep}} u$.

Proof. By strong induction on $|t|$. If $|t| = 1$, then $\mathit{chop}(t\, u) = u$ and thus $t\, u >_{\mathsf{ep}} u$ by E1. If $|t| > 1$, then $\mathit{chop}(t\, u) = \mathit{chop}(t)\, u$, and by the induction hypothesis $\mathit{chop}(t)\, u >_{\mathsf{ep}} u$. Thus $t\, u >_{\mathsf{ep}} u$ by E1. □

Lemma 8. $t\, u >_{\mathsf{ep}} t$.

Proof. By strong induction on $|t|$. Let $t = \xi\, \bar{t}_n$.

If $\xi \in \Sigma$, we apply E3. We have $\bar{t}_n \cdot u \gg_{\mathsf{ep}}^{\xi} \bar{t}_n$ by properties X8 and X10. If $n \neq 0$, we apply the induction hypothesis on $\mathit{chop}(t)$ to obtain $\mathit{chop}(t\, u) = \mathit{chop}(t)\, u >_{\mathsf{ep}} \mathit{chop}(t)$, and we apply E1 to obtain $t\, u >_{\mathsf{ep}} \mathit{chop}(t)$, as required for E3.

If $\xi \in \mathcal{V}$, we apply E4. We have $\bar{t}_n \cdot u \gg_{\mathsf{ep}}^{\mathsf{f}} \bar{t}_n$ for all f by properties X8 and X10. If $n \neq 0$, we apply the induction hypothesis on $\mathit{chop}(t)$ to obtain $\mathit{chop}(t)\, u >_{\mathsf{ep}} \mathit{chop}(t)$. Thus, $\mathit{chop}(t\, u) = \mathit{chop}(t)\, u >_{\mathsf{ep}} \mathit{chop}(t)$ as required for E4. □

Theorem 9 (Subterm Property). *For all subterms s of a term t, we have $t \geq_{\mathsf{ep}} s$.*

Proof. Follows from Lemmas 7 and 8. □

Lemma 10 (Compatibility with Functions). *If $v >_{\text{ep}} u$, then $s\,v >_{\text{ep}} s\,u$.*

Proof. By induction on $|s|$.

Let $s = \zeta\,\bar{s}$. Depending on whether $\zeta \in \Sigma$ or $\zeta \in \mathcal{V}$, we show $s\,v >_{\text{ep}} s\,u$ by applying E3 or E4. By compatibility with tuple contexts (property X6), $v >_{\text{ep}} u$ implies $\bar{s} \cdot v \gg_{\text{ep}}^{\text{f}} \bar{s} \cdot u$ for all $\text{f} \in \Sigma$. Obviously, the tuples $\bar{s} \cdot v$ and $\bar{s} \cdot u$ are not empty. So it remains to show $s\,v >_{\text{ep}} \mathit{chop}(s\,u)$ if $\zeta \in \Sigma$ or $\mathit{chop}(s\,v) >_{\text{ep}} \mathit{chop}(s\,u)$ if $\zeta \in \mathcal{V}$. By E1, it suffices to show $\mathit{chop}(s\,v) >_{\text{ep}} \mathit{chop}(s\,u)$ in both cases.

If $\bar{s} = ()$, then $\mathit{chop}(s\,v) = v >_{\text{ep}} u = \mathit{chop}(s\,u)$ by assumption. Otherwise, $\mathit{chop}(s\,v) = \mathit{chop}(s)\,v >_{\text{ep}} \mathit{chop}(s)\,u = \mathit{chop}(s\,u)$ by the induction hypothesis. □

Lemma 11. *If $t >_{\text{ep}} s$ and $v \geq_{\text{ep}} u$, then $t\,v >_{\text{ep}} s\,u$.*

Proof. By induction on $|t| + |s|$ and a case distinction on how $t >_{\text{ep}} s$ is derived. Let $t = \xi\,\bar{t}_n$ and $s = \zeta\,\bar{s}_m$.

If $t >_{\text{ep}} s$ is derived by E1, then $\mathit{chop}(t) \geq_{\text{ep}} s$. By E1, $t\,v >_{\text{ep}} \mathit{chop}(t\,v) = \mathit{chop}(t)\,v$. So it suffices to show $\mathit{chop}(t)\,v \geq_{\text{ep}} s\,u$. If $\mathit{chop}(t) = s$, this follows from Lemma 10. Otherwise, we have $\mathit{chop}(t) >_{\text{ep}} s$ and hence $\mathit{chop}(t)\,v >_{\text{ep}} s\,u$ holds by the induction hypothesis.

If $t >_{\text{ep}} s$ is derived by E2, then $\xi \succ \zeta$ and either $m = 0$ or $t >_{\text{ep}} \mathit{chop}(s)$. To derive $t\,v >_{\text{ep}} s\,u$ using E2, it remains to show $t\,v >_{\text{ep}} \mathit{chop}(s\,u)$. If $m = 0$, then $\mathit{chop}(s\,u) = u$. Therefore, by the subterm property (Theorem 9), $t\,v >_{\text{ep}} v \geq_{\text{ep}} u = \mathit{chop}(s\,u)$. If $m > 0$, then $t >_{\text{ep}} \mathit{chop}(s)$, and hence by the induction hypothesis, $t\,v >_{\text{ep}} \mathit{chop}(s)\,u = \mathit{chop}(s\,u)$.

If $t >_{\text{ep}} s$ is derived by E3 or E4, we need to show that $\bar{t}_n \gg_{\text{ep}}^{\text{f}} \bar{s}_m$ implies $\bar{t}_n \cdot v \gg_{\text{ep}}^{\text{f}} \bar{s}_m \cdot u$ for all $\text{f} \in \Sigma$. We have $\bar{t}_n \cdot v \gg_{\text{ep}}^{\text{f}} \bar{s}_m \cdot v$ by compatibility with appending (property X9). If $v = u$, we are done. Otherwise, since $\bar{s}_m \cdot v \gg_{\text{ep}}^{\text{f}} \bar{s}_m \cdot u$ by compatibility with tuple contexts (property X6), it follows that $\bar{t}_n \cdot v \gg_{\text{ep}}^{\text{f}} \bar{s}_m \cdot u$ by preservation of transitivity (property X3) and transitivity of $>_{\text{ep}}$ (Theorem 5).

If $t >_{\text{ep}} s$ is derived by E3, we can apply E3 to derive $t\,v >_{\text{ep}} s\,u$. The condition $t\,v >_{\text{ep}} \mathit{chop}(s\,u)$ can be shown as we did for E2 above.

If $t >_{\text{ep}} s$ is derived by E4, we can apply E4 to derive $t\,v >_{\text{ep}} s\,u$. The proof for the condition $\mathit{chop}(t\,v) >_{\text{ep}} \mathit{chop}(s\,u)$ is similar to the argument made for E2 above. □

Theorem 12 (Monotonicity). *If $t >_{\text{ep}} s$, then $u\,t >_{\text{ep}} u\,s$ and $t\,u >_{\text{ep}} s\,u$.*

Proof. By Lemmas 10 and 11. □

Theorem 13 (Embedding Property). *$t \trianglerighteq_{\text{emb}} s$ implies $t \geq_{\text{ep}} s$.*

Proof. By induction on $t \trianglerighteq_{\mathsf{emb}} s$, it suffices to assume that $t \trianglerighteq_{\mathsf{emb}} s$ consists of a single step $t \longrightarrow_{\mathsf{emb}} s$. By Theorems 9 and 12, we then have $t >_{\mathsf{ep}} s$. □

Theorem 14 (Stability under Substitutions). *If $t >_{\mathsf{ep}} s$, then $t\sigma >_{\mathsf{ep}} s\sigma$.*

Proof. By well-founded induction on the multiset $\{|t|, |s|\}$ with respect to the multiset extension [18] of $>$ on $\mathbb{N}$, followed by a case distinction on how $t >_{\mathsf{ep}} s$ is derived. Let $t = \xi\,\bar{t}_n$ and $s = \zeta\,\bar{s}_m$.

If $t >_{\mathsf{ep}} s$ is derived by E1, then $\mathit{chop}(t) \geq_{\mathsf{ep}} s$. By the induction hypothesis, $\mathit{chop}(t)\sigma \geq_{\mathsf{ep}} s\sigma$. Since $t\sigma \longrightarrow_{\mathsf{emb}} \mathit{chop}(t)\sigma$, we have $t\sigma >_{\mathsf{ep}} \mathit{chop}(t)\sigma$ by the embedding property (Theorem 13). Hence, by transitivity $t\sigma >_{\mathsf{ep}} s\sigma$.

If $t >_{\mathsf{ep}} s$ is derived by E2, then $\xi, \zeta \in \Sigma$, $\xi \succ \zeta$, and either $m = 0$ or $t >_{\mathsf{ep}} \mathit{chop}(s)$. We show $t\sigma >_{\mathsf{ep}} s\sigma$ by applying E2. Since $\xi, \zeta \in \Sigma$, the head of $t\sigma$ is ξ, the head of $s\sigma$ is ζ, and the number of arguments of $s\sigma$ is m. Hence, it only remains to show that $t >_{\mathsf{ep}} \mathit{chop}(s)$ implies $t\sigma >_{\mathsf{ep}} \mathit{chop}(s\sigma)$, which follows from the induction hypothesis and from $\mathit{chop}(s)\sigma = \mathit{chop}(s\sigma)$.

If $t >_{\mathsf{ep}} s$ is derived by E3, then $\xi = \zeta \in \Sigma$, $\bar{t}_n \gg_{\mathsf{ep}}^{\zeta} \bar{s}_m$, and either $m = 0$ or $t >_{\mathsf{ep}} \mathit{chop}(s)$. Since $\xi, \zeta \in \Sigma$, the head of $t\sigma$ is ξ, the head of $s\sigma$ is ζ, and $\bar{t}_n\sigma$ and $\bar{s}_m\sigma$ are the respective argument tuples of $t\sigma$ and $s\sigma$. By the induction hypothesis and preservation of stability (property X2) on the set of elements of $\bar{t}_n$ and $\bar{s}_m$, we have $\bar{t}_n\sigma \gg_{\mathsf{ep}}^{\zeta} \bar{s}_m\sigma$. We apply E3 to show $t\sigma >_{\mathsf{ep}} s\sigma$. It remains to show that $t >_{\mathsf{ep}} \mathit{chop}(s)$ implies $t\sigma >_{\mathsf{ep}} \mathit{chop}(s\sigma)$, which follows from the induction hypothesis and from $\mathit{chop}(s)\sigma = \mathit{chop}(s\sigma)$.

If $t >_{\mathsf{ep}} s$ is derived by E4, then $\xi = \zeta \in \mathcal{V}$, $\bar{t}_n \gg_{\mathsf{ep}}^{\mathsf{f}} \bar{s}_m$ for all $\mathsf{f} \in \Sigma$, $n > 0$, and either $m = 0$ or $\mathit{chop}(t) >_{\mathsf{ep}} \mathit{chop}(s)$. We will show that $u\,(\bar{t}_n\sigma) >_{\mathsf{ep}} u\,(\bar{s}_m\sigma)$ for all u with $|u| \leq |\zeta\sigma|$. For $u = \zeta\sigma$, it then follows that $t\sigma >_{\mathsf{ep}} s\sigma$. We show this by induction on $|u|$. We will refer to this induction as the inner induction and to the induction on the multiset $\{|t|, |s|\}$ as the outer induction.

We have to show $u\,(\bar{t}_n\sigma) >_{\mathsf{ep}} u\,(\bar{s}_m\sigma)$. We apply E3 or E4 to do so, depending on whether the head of u is a symbol or a variable. We write $u = \psi\,\bar{u}_r$.

First, we show that $\bar{u}_r \cdot (\bar{t}_n\sigma) \gg_{\mathsf{ep}}^{\mathsf{f}} \bar{u}_r \cdot (\bar{s}_m\sigma)$ for all $\mathsf{f} \in \Sigma$. As above, by the outer induction hypothesis and preservation of stability (property X2) on the set of elements of $\bar{t}_n$ and $\bar{s}_m$, we have $\bar{t}_n\sigma \gg_{\mathsf{ep}}^{\mathsf{f}} \bar{s}_m\sigma$. Then $\bar{u}_r \cdot (\bar{t}_n\sigma) \gg_{\mathsf{ep}}^{\mathsf{f}} \bar{u}_r \cdot (\bar{s}_m\sigma)$ follows by compatibility with prepending (property X8).

If $m = 0$ and $r = 0$, we can apply E3 or E4 directly to show $u\,(\bar{t}_n\sigma) >_{\mathsf{ep}} u\,(\bar{s}_m\sigma)$.

If $r > 0$, then $\mathit{chop}(u\,(\bar{t}_n\sigma)) = \mathit{chop}(u)\,(\bar{t}_n\sigma) >_{\mathsf{ep}} \mathit{chop}(u)\,(\bar{s}_m\sigma) = \mathit{chop}(u\,(\bar{s}_m\sigma))$ by the inner induction hypothesis. If $\psi \in \mathcal{V}$, we can then apply E4 to obtain $u\,(\bar{t}_n\sigma) >_{\mathsf{ep}} u\,(\bar{s}_m\sigma)$. Otherwise, $\psi \in \Sigma$, and we can apply E1 to obtain $u\,(\bar{t}_n\sigma) >_{\mathsf{ep}} \mathit{chop}(u\,(\bar{s}_m\sigma))$ and then E3 to obtain $u\,(\bar{t}_n\sigma) >_{\mathsf{ep}} u\,(\bar{s}_m\sigma)$.

If $m > 0$ and $r = 0$, then we have $chop(t) >_{ep} chop(s)$, $chop(u\ (\bar t_n \sigma)) = chop(t)\sigma$, and $chop(u\ (\bar s_m \sigma)) = chop(s)\sigma$. By the outer induction hypothesis, $chop(t)\sigma >_{ep} chop(s)\sigma$, i.e., $chop(u\ (\bar t_n \sigma)) >_{ep} chop(u\ (\bar s_m \sigma))$. As above, if $\psi \in \mathcal{V}$, we can then apply E4 to obtain $u\ (\bar t_n \sigma) >_{ep} u\ (\bar s_m \sigma)$. Otherwise, $\psi \in \Sigma$, and we can apply E1 to obtain $u\ (\bar t_n \sigma) >_{ep} chop(u\ (\bar s_m \sigma))$ and then E3 to obtain $u\ (\bar t_n \sigma) >_{ep} u\ (\bar s_m \sigma)$.

This concludes the inner and the outer induction. □

Theorem 15 (Ground Totality). *For ground terms t and s, we have $t <_{ep} s$, $t = s$, or $t >_{ep} s$.*

Proof. By well-founded induction on the multiset $\{|t|, |s|\}$ with respect to the multiset extension [18] of $>$ on $\mathbb{N}$. Let $t = \xi\ \bar t_n$ and $s = \zeta\ \bar s_m$. Then $\xi, \zeta \in \Sigma$ because t and s are ground.

If $n > 0$ and $chop(t) \not<_{ep} s$, then by the induction hypothesis $chop(t) \geq_{ep} s$ and hence $t >_{ep} s$ by E1. Thus we can assume that either $n = 0$ or $s >_{ep} chop(t)$. Analogously, we can assume that either $m = 0$ or $t >_{ep} chop(s)$.

If $\xi \succ \zeta$ or $\xi \prec \zeta$, we have $t >_{ep} s$ or $t <_{ep} s$ by E2. Otherwise, we have $\xi = \zeta$ by totality of $\succ$. If either $\bar t \gg^\zeta_{ep} \bar s$ or $\bar t \ll^\zeta_{ep} \bar s$, then we have $t >_{ep} s$ or $t <_{ep} s$ by E3. By the induction hypothesis and preservation of totality (property X7) on the set of elements of $\bar s$ and $\bar t$, if $\bar t \not\gg^\zeta_{ep} \bar s$ and $\bar t \not\ll^\zeta_{ep} \bar s$, then $\bar t = \bar s$ and hence $t = s$. □

Theorem 16 (Well-Foundedness). *The order $>_{ep}$ is well founded.*

Proof. For finite signatures, simplification orders are always well-founded [1, Proposition 6.3.15(ii)]. For infinite signatures, we need to prove well-foundedness. A short proof is to invoke Theorem 5.3 of Middeldorp and Zantema [30]. (Note that their definition of a simplification order differs from mine.) In the Isabelle/HOL formalization, it was more convenient to use the following direct proof.

We suppose that there exists an infinite descending chain $s_0 >_{ep} s_1 >_{ep} \cdots$ and derive a contradiction. We use a minimal counterexample argument [20].

A term s is *bad* if there is an infinite descending $>_{ep}$-chain from s. Other terms are *good*. Without loss of generality, we assume that s_0 has minimal size among all bad terms and that s_{i+1} has minimal size among all bad terms u with $s_i >_{ep} u$.

For each i, let $U_i = \{u \mid s_i \rhd_{emb} u\}$, where $\rhd_{emb}$ is the irreflexive counterpart of $\unrhd_{emb}$. Let $U = \bigcup_i U_i$. All terms in U are good: If there existed a bad $u \in U_0$, then $|s_0| > |u|$, contradicting the minimality of s_0. If there existed a bad $u \in U_{i+1}$ for some i, then $s_i >_{ep} s_{i+1} >_{ep} u$ by the embedding property (Theorem 13), contradicting the minimality of s_{i+1}.

Only conditions E2, E3, and E4 can have been used to derive $s_i >_{ep} s_{i+1}$. If E1 was used, then $chop(s_i) \geq_{ep} s_{i+1} >_{ep} s_{i+2}$. But then there would be an infinite descending chain $\mathit{chop}(s_i) >_{\mathsf{ep}} s_{i+2} >_{\mathsf{ep}} s_{i+3} >_{\mathsf{ep}} \cdots$ from $\mathit{chop}(s_i)$, contradicting the goodness of $\mathit{chop}(s_i) \in U$.

E2 can have been used only finitely many times in the chain since E3 and E4 preserve the head and E2 makes the head smaller with respect to the well-founded relation $\succ$. Hence, there is a number k such that the entire chain $s_k >_{\mathsf{ep}} s_{k+1} >_{\mathsf{ep}} \cdots$ has been derived by E3 and E4. Let $s_i = \zeta\,\bar{u}_i$ (where contrary to our usual convention the indices of $\bar{u}_i$ identify the tuple and do not denote its length). Then we have an infinite chain $\bar{u}_k \gg_{\mathsf{ep}}^{\mathsf{f}} \bar{u}_{k+1} \gg_{\mathsf{ep}}^{\mathsf{f}} \cdots$ for some f. All elements of these tuples are in U because each element of $\bar{u}_i$ is embedded in s_i. However, since all elements of U are good, $>_{\mathsf{ep}}$ is well founded on U. By preservation of well-foundedness (property X5), $\gg_{\mathsf{ep}}^{\mathsf{f}}$ is well founded on U^*, which contradicts the existence of the above $\gg_{\mathsf{ep}}^{\mathsf{f}}$-chain. □

6 Implementation

I implemented EPO in the Zipperposition prover. Zipperposition [14, 15] is an open source[2] superposition-based theorem prover for first- and higher-order logic written in OCaml. In previous work [7], together with colleagues I extended it with refutationally complete modes for λ-free higher order logic, also known as applicative first-order logic. We will focus on the mode that performed best in the evaluation of that paper, the "nonpurifying intensional variant". It is designed to deal with nonmonotonic orders such as λfRPO, but falls back to a simpler calculus with monotonic orders, such as λfKBO or EPO.

The pseudocode of the prototype implementation of EPO is given in Figure 1. As usual in superposition provers, the procedure compares two terms in both directions, yielding one of the answers GreaterThan, Equal, LessThan, or Incomparable. When the pseudocode refers to $>_{\mathsf{ep}}$, $\geq_{\mathsf{ep}}$, and $\gg_{\mathsf{ep}}^{\mathsf{f}}$, this is to be interpreted in terms of the function epo. The syntax '$\xi\,\bar{t}_n$ as t' in the arguments of function definitions means that t denotes the entire term, ξ denotes its head, and $\bar{t}_n$ denotes its arguments.

It is crucial to the performance of this implementation to use memoization in the form of a cache on the function epo. For example, to compute that $\mathsf{f}^m\,x \not\leq_{\mathsf{ep}} \mathsf{f}^n\,y$ for $m \leq n$, we need at least 4^m calls to epo if the cache is inactive. With a cache however, only $(m+1)(n+1)$ of these calls to epo have to be computed; the other return values can be found in the cache. More generally, the following lemma holds:

Lemma 17. *To calculate the order of two terms t and s, the pseudocode in Figure 1 needs at most* $\mathrm{depth}(t) \cdot \mathrm{depth}(s) \cdot |t| \cdot |s|$ *distinct calls to* epo. *Here, the depth of a term $\zeta\,\bar{u}_m$ is 1 if $m = 0$ and $\max_{u \in \bar{u}}(\mathrm{depth}(u)) + 1$ otherwise.*

---

[2] https://github.com/sneeuwballen/zipperposition

```
epo(ξ t̄ₙ as t, ζ s̄ₘ as s) =
  if t = s then Equal
  elif t ∈ 𝒱 and s ∈ 𝒱 then Incomparable
  elif t ∈ 𝒱 then (if t occurs in s then LessThan else Incomparable)
  elif s ∈ 𝒱 then (if s occurs in t then GreaterThan else Incomparable)
  else
    if ξ ≻ ζ then check_{E2,E3}(t, s)
    elif ξ ≺ ζ then check^{inv}_{E2,E3}(t, s)
    elif ξ = ζ and ζ ∈ Σ then
      if t̄ₙ ≫^ζ_{ep} s̄ₘ then check_{E2,E3}(t, s)
      elif t̄ₙ ≪^ζ_{ep} s̄ₘ then check^{inv}_{E2,E3}(t, s)
      else check_{E1}(t, s)
    elif ξ = ζ and ζ ∈ 𝒱 then
      if t̄ₙ ≫^f_{ep} s̄ₘ for all f ∈ Σ and n > 0 then check_{E4}(t, s)
      elif t̄ₙ ≪^f_{ep} s̄ₘ for all f ∈ Σ and m > 0 then check^{inv}_{E4}(t, s)
      else check_{E1}(t, s)
    else check_{E1}(t, s)

check_{E1}(ξ t̄ₙ as t, ζ s̄ₘ as s) =
  if n > 0 and chop(t) ≥_{ep} s then GreaterThan
  elif m > 0 and t ≤_{ep} chop(s) then LessThan
  else Incomparable

check_{E2,E3}(ξ t̄ₙ as t, ζ s̄ₘ as s) =
  if m = 0 or t >_{ep} chop(s) then GreaterThan else check_{E1}(t, s)

check^{inv}_{E2,E3}(ξ t̄ₙ as t, ζ s̄ₘ as s) =
  if n = 0 or chop(t) <_{ep} s then LessThan else check_{E1}(t, s)

check_{E4}(ξ t̄ₙ as t, ζ s̄ₘ as s) =
  if m = 0 or chop(t) >_{ep} chop(s) then GreaterThan else check_{E1}(t, s)

check^{inv}_{E4}(ξ t̄ₙ as t, ζ s̄ₘ as s) =
  if n = 0 or chop(t) <_{ep} chop(s) then LessThan else check_{E1}(t, s)
```

Figure 1: Pseudocode of the EPO implementation

Proof. We define a set S_t that overapproximates the set of all embeddings of t that may be involved in computing the order of t with some other term.

To this end, let $\rhd_{\mathsf{arg}}$ be the relation defined by $\zeta\,\bar{u}_n \rhd_{\mathsf{arg}} u_i$ for all terms $\zeta\,\bar{u}_n$ and all i. Let $\rhd_{\mathsf{chop}}$ be the relation defined by $\zeta\,\bar{u}_n \rhd_{\mathsf{chop}} \mathit{chop}(\zeta\,\bar{u}_n)$ for all terms $\zeta\,\bar{u}_n$ with $n > 0$. Finally, let S_t be the set of all terms u such that $t\,(\rhd_{\mathsf{arg}} \cup \rhd_{\mathsf{chop}})^*\,u$. In other words, S_t is inductively defined as follows: Let $t \in S_t$. For any term $\zeta\,\bar{u}_n \in S_t$, let $\mathit{chop}(\zeta\,\bar{u}_n) \in S_t$ and $u_i \in S_t$ for all i.

Inspecting the pseudocode, it is obvious that S_t and S_s together overapproximate all terms that are involved in computing the order for the two terms t and s.

In a derivation of $(\rhd_{\mathsf{arg}} \cup \rhd_{\mathsf{chop}})^*$, any $\rhd_{\mathsf{chop}}$ step before a $\rhd_{\mathsf{arg}}$ step can be eliminated. More precisely, we show that $(\rhd_{\mathsf{arg}} \cup \rhd_{\mathsf{chop}})^* = (\rhd_{\mathsf{arg}}^* \circ \rhd_{\mathsf{chop}}^*)$ by proving that $(\rhd_{\mathsf{chop}} \circ \rhd_{\mathsf{arg}}) \subseteq (\rhd_{\mathsf{arg}}^*)$. We assume that $w \rhd_{\mathsf{chop}} v \rhd_{\mathsf{arg}} u$ for some terms w, v, and u. Let $w = \zeta\,\bar{w}_n$. Then $v = \mathit{chop}(\zeta\,\bar{w}_n) = w_1\,w_2\,\ldots\,w_n$. Let $w_1 = \xi\,\bar{v}_n$. Then $v = \xi\,\bar{v}_n\,w_2\,\ldots\,w_n$. Hence $u \in \bar{v}_n$ or $u \in \{w_2, \ldots, w_n\}$. In the first case, we have $w \rhd_{\mathsf{arg}} w_1 \rhd_{\mathsf{arg}} u$; In the second case $w \rhd_{\mathsf{arg}} u$. Either way, $w\,(\rhd_{\mathsf{arg}}^*)\,u$, which is what we needed to show.

Hence, S_t is the set of all terms v such that $t\,(\rhd_{\mathsf{arg}}^* \circ \rhd_{\mathsf{chop}}^*)\,v$. Therefore, we can overapproximate the size of S_t as follows:

$$|S_t| \leq \sum_{u \in \mathcal{T},\, t \rhd_{\mathsf{arg}}^* u} |\{v \mid u \rhd_{\mathsf{chop}}^* v\}| \leq \sum_{u \in \mathcal{T},\, t \rhd_{\mathsf{arg}}^* u} |u| \leq \mathrm{depth}(t) \cdot |t|$$

The last inequality holds because for any number of steps k,

$$\sum_{u \in \mathcal{T},\, t \rhd_{\mathsf{arg}}^k u} |u| \leq |t|$$

and the number of $\rhd_{\mathsf{arg}}$ steps from t is bounded by $\mathrm{depth}(t)$.

Since S_t and S_s together overapproximate all terms that are involved in computing the order for the two terms t and s, we can overapproximate the number of distinct calls to epo by $|S_t \times S_s| = |S_t| \cdot |S_s| \leq \mathrm{depth}(t) \cdot |t| \cdot \mathrm{depth}(s) \cdot |s|$. □

We can use this lemma to derive the computational complexity of epo. The following theorem is stated only for the length-lexicographic extension operators since other extension operators may have a higher computational complexity.

Theorem 18. *For each* $\mathsf{f} \in \Sigma$, *let* $> \mapsto \gg^{\mathsf{f}}$ *be either the left-to-right or the right-to-left length-lexicographic extension operator. For terms t and s, the computational complexity of* $\mathsf{epo}(t, s)$ *as given in Figure 1 is* $O(\mathrm{depth}(t) \cdot \mathrm{depth}(s) \cdot |t| \cdot |s| \cdot (|t| + |s|))$ *if recursive calls are cached.*

Proof. Let $R(t,s)$ be the set of term pairs (v,u), for which $\mathsf{epo}(t,s)$ triggers directly or indirectly a call to $\mathsf{epo}(v,u)$. Let $C(v,u)$ be the complexity of $\mathsf{epo}(v,u)$ assuming $O(|v|+|u|)$ for all recursive calls. Then the computational complexity of $\mathsf{epo}(t,s)$ is

$$O\left(\sum_{(v,u)\in R(t,s)} C(v,u)\right) \qquad (*)$$

We assume $O(|v|+|u|)$ for the recursive calls in the definition of $C(v,u)$ because each recursive call is either the first one for this argument pair and therefore counted by another summand of the sum above, or it is not the first one for this argument pair and can therefore be retrieved from the cache in $O(|v|+|u|)$. (Zipperposition can retrieve the result from the cache even in constant time because it uses hash consing for terms.)

To determine $C(v,u)$, we analyze the implementation in Figure 1, assuming that all recursive calls are linear. Retrieving a result from the cache, searching for occurrences of a given variable in a term, computing *chop*, counting the number of arguments of a term, and iterating through the arguments for the length-lexicographic comparison are $O(|v|+|u|)$. All other operations are $O(1)$. Hence, $C(v,u)$ is $O(|v|+|u|)$. Since the term sizes do not increase in recursive calls, $C(v,u)$ is also $O(|t|+|s|)$ for all $(v,u) \in R(t,s)$. By Lemma 17, $|R(t,s)| \leq \mathrm{depth}(t)\cdot\mathrm{depth}(s)\cdot|t|\cdot|s|$. Hence, by (*), the computational complexity of $\mathsf{epo}(t,s)$ is $O(\mathrm{depth}(t)\cdot\mathrm{depth}(s)\cdot|t|\cdot|s|\cdot(|t|+|s|))$. □

Compared with first-order KBO or RPO, this is rather slow. Löchner [27, 28] showed that, with a lexicographic extension, KBO can be computed in $O(|t|+|s|)$ and RPO in $O(|t|\cdot|s|)$. RPO can be implemented so efficiently because the computation of the lexicographic order of the arguments, i.e., computing $\bar{t}_n \gg_{\mathsf{ep}}^{\zeta} \bar{s}_m$, can be merged with testing other conditions, i.e., the condition corresponding to $\mathsf{check}_{\mathsf{E2,E3}}(t,s)$. It is an open question whether a similar optimization is possible for EPO, although it is definitely not as straightforward as for RPO.

7 Evaluation

The following evaluation compares the prototype implementation of EPO with other orders in Zipperposition. It was performed with a CPU time limit of 300 s on StarExec Iowa nodes equipped with Intel Xeon E5-2609 0 CPUs clocked at 2.40 GHz. The raw evaluation results are available online and reproducible.[3]

---
[3] https://doi.org/10.5281/zenodo.3992684

From the TPTP [35], 665 higher-order problems in THF format were used, containing both monomorphic and polymorphic problems and excluding problems that contain arithmetic, tuples, the $distinct predicate, or the $ite symbol, as well as problems whose clausal normal form falls outside the λ-free fragment.

The Sledgehammer (SH) benchmarks, corresponding to the Isabelle's Judgment Day problems [12], were regenerated to target λ-free higher-order logic, encoding λ-expressions as λ-lifted supercombinators [29]. The SH benchmarks comprise 1253 problems, each including 256 Isabelle facts.

Besides EPO, I evaluate λfRPO, λfKBO, and their applicative counterparts (appRPO, appKBO). Each of the orders were used twice, once using the left-to-right length-lexicographic extension (LTR) and once using the right-to-left length-lexicographic extension (RTL) for all symbols. In principle, EPO also allows for different extension operators for different symbols, but it is unclear how to design appropriate heuristics. For all orders, I use the inverse frequency of symbols as precedence. On first-order benchmarks, λfRPO and λfKBO coincide with first-order RPO and KBO. The calculus used for EPO, λfRPO, and λfKBO is the intensional nonpurifying variant of the calculus described in my earlier work [7]. For the monotonic orders EPO and λfKBO, the calculus degrades to essentially first-order superposition, with the addition of an argument congruence rule that adds arguments of partially applied functions. In the case of the nonmonotonic order λfRPO, the calculus performs additional superposition inferences onto variables to remain complete, which is why we would generally expect a better performance with monotonic orders. To evaluate the applicative counterparts appKBO and appRPO, I apply the applicative encoding to the given problem directly after the clausal normal form transformation and use first-order KBO and RPO, respectively, on the resulting problem. The results for these last two orders are therefore to be interpreted with care because the applicative encoding also influences various unrelated heuristics in Zipperposition.

Figure 2 displays the number of problems found to be satisfiable (#sat), the number of problems found to be unsatisfiable (#uns), the average CPU time per problem (∅tim), the average percentage of the CPU time used to compute order comparisons (%ord), and the average number of clauses produced during a run (∅cla). When computing the three averages, satisfiable problems and problems that at least one of the ten configurations failed to solve within the time limit were excluded.

From first-order provers, it is well known that KBO generally outperforms RPO. In the #uns columns, we observe the same effect. In the present setting, the advantage of λfKBO is possibly even greater because the calculus performs inferences onto variables with RPO. Although these additional superposition inferences are not performed when using EPO, the #uns results for EPO are worse than λfRPO and

		LTR					RTL				
		#sat	#uns	∅tim	%ord	∅cla	#sat	#uns	∅tim	%ord	∅cla
TPTP	EPO	120	463	1.3	6.9	2155	120	462	1.2	6.8	2163
	λfRPO	119	472	0.3	0.9	1196	119	471	0.3	1.0	1171
	λfKBO	121	474	0.1	1.6	430	121	473	0.2	1.6	600
	appRPO	138	472	0.6	1.1	749	123	472	1.6	2.0	1489
	appKBO	122	476	0.1	1.9	306	122	476	0.3	2.0	462
SH	EPO	1	509	2.6	23.5	6356	1	505	3.1	23.2	6251
	λfRPO	1	550	1.6	4.7	7130	1	549	2.4	4.8	8612
	λfKBO	1	594	1.6	8.8	9206	1	590	1.3	8.7	6949
	appRPO	1	481	13.3	8.1	26346	1	462	17.9	16.3	28897
	appKBO	1	502	10.6	11.3	25236	1	502	10.9	11.6	26202

Figure 2: Evaluation

λfKBO. The %ord columns reveal that this is probably because EPO takes considerably more time to compute. I hypothesized that a second reason could be that generally more term pairs are incomparable under EPO and thus more inferences need to be performed and more clauses are produced. Although the numbers in the ∅cla column on the TPTP benchmark set confirm this hypothesis, the corresponding numbers on the SH benchmark set contradict it because on those benchmarks, EPO is actually producing the least amount of clauses.

The raw data indicate that despite the poor performance of λfRPO and EPO these orders may be useful in a portfolio prover. The λfRPO configurations can solve 16 problems that neither of the λfKBO configurations can solve. The EPO configurations can solve 11 problems that neither of the λfRPO configurations can solve, 12 problems that neither of the λfKBO configurations can solve, 51 problems that neither of the appRPO configurations can solve, 66 problems that neither of the appKBO configurations can solve, and 4 problems that no other configuration can solve. Most of the problems where EPO outperforms other orders are in the SH benchmark set. Overall, λfRPO is preferable over EPO if one is willing to face the complications of a nonmonotonic order in theory and in implementation.

The direction (LTR or RTL) of the length-lexicographic extension does not have a large impact. For λfKBO and appKBO, this is to be expected since the lexicographic comparison comes into play only when weights are equal. For EPO, the advantage of RTL suggested in Section 4.2 is not corroborated by the evaluation. Only with appRPO, LTR performs better than RTL. This might be because LTR tends to put more importance to the symbols that were at the heads of terms before the applicative encoding, yielding a better measure of the complexity of a term.

8 Discussion and related work

I presented a ground-total and well-founded simplification order for λ-free higher-order terms resembling RPO. In first-order logic, KBO generally outperforms RPO, but RPO with well-chosen parameters behaves better than KBO on many examples. In λ-free higher-order logic, the situation appears to be similar. However, RPO cannot be easily used for superposition in this logic if we want the calculus to remain complete because the natural generalization [9] is nonmonotonic. If one wants to avoid the complications of nonmonotonic orders, EPO seems to be a good replacement to fill the role of RPO in λ-free higher-order logic. Otherwise, calculi specialized to deal with nonmonotonic orders such as λfRPO [7, 11] are the better choice.

The literature contains several other variants of RPO targeting the more difficult problem of providing useful orders for full higher-order terms with λ-abstractions: Lifantsev and Bachmair's lexicographic path-order on λ-free higher-order terms [26], Jouannaud and Rubio's higher-order RPO (HORPO) [22], Kop and Van Raamsdonk's iterative HORPO [25], the HORPO extension with polynomial interpretation orders by Bofill et al. [11], and the computability path order by Blanqui et al. [10]. However, these orders all lack ground-totality and, except for Lifantsev and Bachmair's order, the subterm property for terms of different types.

Goubault-Larrecq [21] and Dershowitz [17] provide general frameworks to prove well-foundedness of RPO-like orders. I have considered using them, but determined that they would not reduce the overall complexity of my proofs because establishing that these frameworks apply to EPO is not trivial, and EPO's well-foundedness is not the most difficult property to establish. In fact, the subterm property and stability under substitutions are the ones that are difficult to show. Goubault-Larrecq's framework offers a lemma to prove stability under substitutions, but unfortunately it is limited to first-order logic.

To explore different candidate definitions for EPO, I formalized my ideas early on in Isabelle/HOL [31]. This allowed me to keep track of changes in the definition and how they influence the properties and their proofs more easily. To find examples explaining why certain properties do not hold for some tentative definitions of EPO, Lazy SmallCheck [33] was of great help. For instance, it was Lazy SmallCheck that found the example $x \mathrel{\mathsf{f}} \mathrel{\mathsf{f}} >'_{\mathsf{ep}} x \, x$ versus $\mathsf{f} \, y \mathrel{\mathsf{f}} \mathrel{\mathsf{f}} \not>'_{\mathsf{ep}} \mathsf{f} \, y \, (\mathsf{f} \, y)$ mentioned in Section 4.

In future work, I would like to investigate whether the computation of EPO can be optimized further. To put EPO to use in practice, implementing it in E prover

[34] would be a good target because E's λ-free higher-order mode is designed for ground-total simplification orders and its calculus is more efficient for those than Zipperposition's by circumventing the argument congruence rule.

Acknowledgments

I am grateful to the maintainers of StarExec Iowa for letting me use their service; to Jasmin Blanchette for his excellent supervision and for providing benchmarks; to Ahmed Bhayat, Simon Cruanes, Wan Fokkink, Carsten Fuhs, and Petar Vukmirović for the stimulating discussions and the constructive feedback; to Jürgen Giesl and Femke van Raamsdonk for pointers to the literature; and to the anonymous reviewers for suggesting many improvements to this text. My research has received funding from the European Research Council (ERC) under the European Union's Horizon 2020 research and innovation program (grant agreement No. 713999, Matryoshka). It has also been funded by a Chinese Academy of Sciences President's International Fellowship for Postdoctoral Researchers (grant No. 2021PT0015).

References

[1] Franz Baader and Tobias Nipkow. *Term Rewriting and All That*. Cambridge University Press, 1998.

[2] Leo Bachmair and Harald Ganzinger. Rewrite-based equational theorem proving with selection and simplification. *J. Log. Comput.*, 4(3):217–247, 1994.

[3] Heiko Becker, Jasmin Christian Blanchette, Uwe Waldmann, and Daniel Wand. A transfinite Knuth–Bendix order for lambda-free higher-order terms. In Leonardo de Moura, editor, *CADE-26*, volume 10395 of *LNCS*, pages 432–453. Springer, 2017.

[4] Alexander Bentkamp. Formalization of the embedding path order for lambda-free higher-order terms. *Archive of Formal Proofs*, 2018. http://isa-afp.org/entries/Lambda_Free_EPO.html.

[5] Alexander Bentkamp. *Superposition for Higher-Order Logic*. PhD thesis, Vrije Universiteit Amsterdam, 2021.

[6] Alexander Bentkamp, Jasmin Blanchette, Simon Cruanes, and Uwe Waldmann. Superposition for lambda-free higher-order logic. Accepted in *Log. Meth. Comput. Sci.* Preprint at https://arxiv.org/abs/2005.02094v2 (2020). arXiv:2005.02094.

[7] Alexander Bentkamp, Jasmin Christian Blanchette, Simon Cruanes, and Uwe Waldmann. Superposition for lambda-free higher-order logic. In Didier Galmiche, Stephan Schulz, and Roberto Sebastiani, editors, *IJCAR 2018*, volume 10900 of *LNCS*, pages 28–46. Springer, 2018.

[8] Jasmin Christian Blanchette, Uwe Waldmann, and Daniel Wand. A lambda-free higher-order recursive path order. Tech. report, http://people.mpi-inf.mpg.de/~jblanche/lambda_free_rpo_rep.pdf, 2016.

[9] Jasmin Christian Blanchette, Uwe Waldmann, and Daniel Wand. A lambda-free higher-order recursive path order. In Javier Esparza and Andrzej S. Murawski, editors, *FOSSACS 2017*, volume 10203 of *LNCS*, pages 461–479. Springer, 2017.

[10] Frédéric Blanqui, Jean-Pierre Jouannaud, and Albert Rubio. The computability path ordering. *Log. Meth. Comput. Sci.*, 11(4), 2015.

[11] Miquel Bofill, Cristina Borralleras, Enric Rodríguez-Carbonell, and Albert Rubio. The recursive path and polynomial ordering for first-order and higher-order terms. *J. Log. Comput.*, 23(1):263–305, 2013.

[12] Sascha Böhme and Tobias Nipkow. Sledgehammer: Judgement Day. In Jürgen Giesl and Reiner Hähnle, editors, *IJCAR 2010*, volume 6173 of *LNCS*, pages 107–121. Springer, 2010.

[13] Iliano Cervesato and Frank Pfenning. A linear spine calculus. *J. Log. Comput.*, 13(5):639–688, 2003.

[14] Simon Cruanes. *Extending Superposition with Integer Arithmetic, Structural Induction, and Beyond*. Ph.D. thesis, École polytechnique, 2015.

[15] Simon Cruanes. Superposition with structural induction. In Clare Dixon and Marcelo Finger, editors, *FroCoS 2017*, volume 10483 of *LNCS*, pages 172–188. Springer, 2017.

[16] Nachum Dershowitz. Orderings for term-rewriting systems. *Theor. Comput. Sci.*, 17:279–301, 1982.

[17] Nachum Dershowitz. Jumping and escaping: Modular termination and the abstract path ordering. *Theor. Comput. Sci.*, 464:35–47, 2012.

[18] Nachum Dershowitz and Zohar Manna. Proving termination with multiset orderings. *Commun. ACM*, 22(8):465–476, 1979.

[19] Naohi Eguchi. A lexicographic path order with slow growing derivation bounds. *Math. Log. Q.*, 55(2):212–224, 2009.

[20] Maria C. F. Ferreira and Hans Zantema. Well-foundedness of term orderings. In Nachum Dershowitz and Naomi Lindenstrauss, editors, *CTRS-94*, volume 968 of *LNCS*, pages 106–123. Springer, 1994.

[21] Jean Goubault-Larrecq. Well-founded recursive relations. In Laurent Fribourg, editor, *Computer Science Logic (CSL 2001)*, volume 2142 of *LNCS*, pages 484–497. Springer, 2001.

[22] Jean-Pierre Jouannaud and Albert Rubio. Polymorphic higher-order recursive path orderings. *J. ACM*, 54(1):2:1–2:48, 2007.

[23] Richard Kennaway, Jan Willem Klop, M. Ronan Sleep, and Fer-Jan de Vries. Comparing curried and uncurried rewriting. *J. Symb. Comput.*, 21(1):15–39, 1996.

[24] D. E. Knuth and P. B. Bendix. Simple word problems in universal algebras. In J. Leech, editor, *Computational Problems in Abstract Algebra*, pages 263–297. Pergamon Press, 1970.

[25] Cynthia Kop and Femke van Raamsdonk. A higher-order iterative path ordering. In Iliano Cervesato, Helmut Veith, and Andrei Voronkov, editors, *LPAR 2008*, volume 5330 of *LNCS*, pages 697–711. Springer, 2008.

[26] Maxim Lifantsev and Leo Bachmair. An LPO-based termination ordering for higher-order terms without λ-abstraction. In Jim Grundy and Malcolm C. Newey, editors, *TPHOLs '98*, volume 1479 of *LNCS*, pages 277–293. Springer, 1998.

[27] Bernd Löchner. Things to know when implementing KBO. *J. Autom. Reason.*, 36(4):289–310, 2006.

[28] Bernd Löchner. Things to know when implementing LPO. *Internat. J. Artificial Intelligence Tools*, 15(1):53–80, 2006.

[29] Jia Meng and Lawrence C. Paulson. Translating higher-order clauses to first-order clauses. *J. Autom. Reason.*, 40(1):35–60, 2008.

[30] Aart Middeldorp and Hans Zantema. Simple termination of rewrite systems. *Theor. Comput. Sci.*, 175(1):127–158, 1997.

[31] Tobias Nipkow, Lawrence C. Paulson, and Markus Wenzel. *Isabelle/HOL - A Proof Assistant for Higher-Order Logic*, volume 2283 of *LNCS*. Springer, 2002.

[32] Lawrence C. Paulson and Jasmin Christian Blanchette. Three years of experience with Sledgehammer, a practical link between automatic and interactive theorem provers. In Geoff Sutcliffe, Stephan Schulz, and Eugenia Ternovska, editors, *IWIL-2010*, volume 2 of *EPiC*, pages 1–11. EasyChair, 2012.

[33] Jason S. Reich, Matthew Naylor, and Colin Runciman. Advances in Lazy SmallCheck. In Ralf Hinze, editor, *IFL*, volume 8241 of *LNCS*, pages 53–70. Springer, 2012.

[34] Stephan Schulz, Simon Cruanes, and Petar Vukmirović. Faster, higher, stronger: E 2.3. In Pascal Fontaine, editor, *CADE-27*, volume 11716 of *LNCS*, pages 495–507. Springer, 2019.

[35] G. Sutcliffe. The TPTP Problem Library and Associated Infrastructure. From CNF to TH0, TPTP v6.4.0. *J. Autom. Reason.*, 59(4):483–502, 2017.

[36] Petar Vukmirović, Jasmin Blanchette, Simon Cruanes, and Stephan Schulz. Extending a brainiac prover to lambda-free higher-order logic. In Tomás Vojnar and Lijun Zhang, editors, *TACAS 2019*, volume 11427 of *LNCS*, pages 192–210. Springer, 2019.

[37] Hans Zantema. Termination. In Marc Bezem, Jan Willem Klop, and Roel de Vrijer, editors, *Term Rewriting Systems*, volume 55 of *Cambridge Tracts in Theoretical Computer Science*, pages 181–259. Cambridge University Press, 2003.

Entailment, Transmission of Truth, and Minimality

Andrzej Wiśniewski
Department of Logic and Cognitive Science
Adam Mickiewicz University in Poznań
Poland
`Andrzej.Wisniewski@amu.edu.pl`

Abstract

Two concepts of single- and multiple-conclusion entailment, based on the idea of minimality, are introduced and studied. The analysis is performed at the propositional level. As for consistent sets of premises, ranges of the entailments defined, dubbed "strong", equal ranges of their classical counterparts. Yet, strong entailments are non-Tarskian. In particular, they are not monotone, but, at the same time, have some intuitively plausible properties which their standard counterparts lack. A proof-theoretic account of minimally inconsistent sets and thus, indirectly, of strong entailments is provided. Some applications of the introduced concepts, pertaining to belief revision and argument analysis, are discussed.

Keywords: entailment, multiple-conclusion entailment, non-monotonic logic, minimally inconsistent sets, contraction, argument analysis

1 Introduction

1.1 Single-Conclusion Entailment

The idea of *transmission of truth* underlies the intuitive concept of entailment. According to the idea, entailment is akin to an input-output device which, when fed with truth at the input, gives truth at the output. The input need not consist of truths, but *if* it does, it transforms into a true output. Similarly, *if* the premises are all true, any conclusion entailed by them must be true, although the truth of premises is not a necessary condition for entailment to hold. Or, to put it differently, the *hypothetical* truth of premises warrants the truth of an entailed conclusion.[1]

Logicians operate with well-formed formulas (*wffs* for short) of formalized languages and conceptualize entailment as a semantic relation between sets of wffs and single wffs. At the same time they tend to understand the "if" above in the sense of material conditional. Yet, since a material conditional with false antecedent is true irrespective of the logical value of the consequent, as a consequence one gets:

(I) *a set of wffs which cannot be simultaneously true, i.e. an inconsistent set, entails every wff.*

Moreover, a material conditional with true consequent is true irrespective of the logical value of the antecedent, and hence:

(II) *a logically valid wff is entailed by any set of wffs.*

Both (I) and (II) are a kind of by-products and we got accustomed to live with them. But (I) as well as (II) seem to contravene the intuitive idea of transmission of truth. To say that "truth is transmitted" seems to presuppose that it *can* occur at the input and that it *need not* occur at the output.

Another drawback of the received view is this. To say that the hypothetical truth of sentences in a set X warrants the truth of a sentence B seems to presuppose that the hypothetical truth of *all* the sentences in X contributes to the hypothetical truth of B. Entailment intuitively construed is a kind of semantic entrenchment of an entailed sentence in a set of sentences that entails it: a set of sentences X that entails a sentence B comprises neither less nor more sentences than those the hypothetical truth of which, jointly, warrants the truth of B. On the other hand, entailment defined in the usual way, by using, inter alia, the material "if", is monotone:

(M) *a wff B entailed by a set of wffs X is entailed by any superset of X as well*

and hence the wff B is also entailed by sets of wffs which contain elements that are irrelevant with regard to the transmission of truth and/or the semantic entrenchment effect(s): their hypothetical truth do not contribute in any way to the truth of B.

1.2 Multiple-Conclusion Entailment

The concept of entailment is sometimes generalized to the concept of multiple-conclusion entailment (*mc-entailment* for short). Mc-entailment is a semantic relation between sets of wffs, where an entailed set is allowed to contain more than one

---

[1] The latter statement can be explicated as: "If the truth-conditions of all the premises are met, an entailed conclusion is true as well."

element. The underlying idea is: an mc-entailed set must contain at least one true wff *if* the respective mc-entailing set consists of truths. Or, to put it differently, the hypothetical truth of all the wffs in an mc-entailing set warrants the existence of a true wff in the mc-entailed set.[2]

Mc-entailment can hold for trivial reasons: X mc-entails Y because X single-conclusion entails (*sc-entails* for short) at least one wff in Y. But mc-entailment can also hold non-trivially: it happens that a set of wffs, X, mc-entails a set of wffs, Y, although X does not sc-entail any wff in Y. For instance (taking Classical Propositional Logic as the basis), the truth of all the wffs in the set $X = \{p \to q \vee r, p\}$ warrants the existence of a true wff in the set $Y = \{q, r\}$, but neither q nor r is sc-entailed by X or, to put it differently, the hypothetical truth of the wffs in X guarantees that at least one of: q, r, is true, but warrants neither the truth of q nor the truth of r.

The concept of mc-entailment is more general than that of sc-entailment. One can always define sc-entailment as mc-entailment of a singleton set. However, it is not the case that mc-entailment can always be defined in terms of sc-entailment.[3]

One of the ways of thinking of entailed non-singleton *sets* is to construe them as items effectively *delimiting* search spaces: a set of wffs Y entailed by a set of wffs X is a minimal set that comprises wffs among which a truth must lie if the wffs in X are all true. "Minimal" means here "no proper subset of Y behaves analogously w.r.t. X." Another way of thinking about an entailed set is to construe it as characterizing the *relevant cases* to be considered, for if X mc-entails Y and each wff in Y sc-entails a wff B, the wff B is sc-entailed by X as well. However, the standard concept of mc-entailment is too broad to reflect the above ideas. This is due to the fact that mc-entailment is *right-monotone*.

(RM) *if a set of wffs X mc-entails a set of wffs, Y, then X mc-entails any superset of Y as well.*

Observe also that mc-entailment explicated by means of the material "if" suffers

---

[2]It is sometimes claimed that the concept of mc-entailment originates from [5] due to his introduction of sequents with sequences of wffs in the succedents. The semantic concept of mc-entailment was explicitly introduced in [4] under the heading "involution." Its syntactic counterpart, mc-consequence, was incorporated into the general theory of logical calculi by [19]. The first monograph devoted to mc-consequence and related concepts (multiple-conclusion calculus, multiple-conclusion rules, etc.) was [20].

[3]Philosophers tend to think that mc-entailment between sets X and Y is nothing more than sc-entailment of a disjunction of all the formulas in Y from a conjunction of all the formulas in X. However, this is not always so; whether this is right depends both on the richness of syntax (the presence/lack of infinite disjunctions and conjunctions) and semantics (the classical/nonclassical meanings of disjunction and conjunction). For details and counterexamples see [21].

similar drawbacks to those of sc-entailment explicated in this way:

(I') *any set of wffs is mc-entailed by an inconsistent set of wffs*, and

(II') *a set of wffs that contains a logically valid wff is mc-entailed by any set of wffs*.

Moreover, mc-entailment is *left-monotone*, that is:

(LM) *a set of wffs Y which is mc-entailed by a set of wffs X is also mc-entailed by any superset of X.*

Hence there exist mc-entailing sets of wffs which contain, inter alia, wffs that are semantically irrelevant to the corresponding mc-entailed sets. For example, $\{s, p \to q \vee r, p\}$ mc-entails $\{q, r\}$, while the hypothetical truth of s is completely irrelevant to the occurrence of truth in $\{q, r\}$.

1.3 Aims

In this paper we introduce and examine a concept of multiple-conclusion entailment, which we dub "strong multiple-conclusion entailment." Formally, strong mc-entailment is a subrelation of mc-entailment. We define strong mc-entailment in a way which allows us to avoid the drawbacks (I') and (II') indicated above. Moreover, strong mc-entailment is neither left-monotone nor right-monotone. As a by-product one gets a concept of single-conclusion entailment which, in turn, is free of the drawbacks pointed out at the beginning of this paper. We coin the concept "strong single-conclusion entailment." This concept of sc-entailment is analysed in the paper as well.

For simplicity, the analysis is pursued at the propositional level.

The paper is organized as follows.

Section 2 introduces the logical apparatus needed.

Section 3 is devoted to strong mc-entailment. Its definition is proposed and the adequacy issue is addressed therein. The consecutive subsections include theorems and corollaries characterizing basic properties of strong mc-entailment. In particular, it is shown that strong mc-entailment between non-empty sets of wffs involves only such sets which comprise contingent (i.e. neither valid nor inconsistent) wffs. It occurs that, as long as the underlying logic has the compactness property, strong mc-entailment holds only between finite sets of wffs. Strong mc-entailment exhibits the variable-sharing property. The relation between strong mc-entailment and minimally inconsistent sets is examined. A deduction theorem for strong mc-entailment is provided.

An analysis of strong sc-entailment is presented in Section 4. We define it as strong mc-entailment of a singleton set. Basic properties of strong sc-entailment are analysed. It is shown that the relation between strong mc- and sc-entailments does not fit the "a conjunction yields a disjunction" pattern.

In Section 5 we compare strong sc- and mc-entailment with their classical counterparts. We show that classical sc-entailment from a consistent set of wffs boils down to strong sc-entailment from a finite subset of the set: any wff classically sc-entailed by a consistent set of wffs is also strongly sc-entailed by a finite subset of the set. An analogous result for strong mc-entailment is proven as well. Section 5 includes also some comparative remarks on strong entailments and accounts of entailment proposed in relevance logics and connective logics.

Section 6 provides a proof-theoretic account of minimally inconsistent sets and thus, indirectly, of strong mc- and sc-entailments. Proofs of soundness and completeness of the proposed calculus are given in the Appendix.

Section 7 is devoted to some conceptual applications of the results presented in the previous sections.

Section 8 discusses the issue of transferability of the results concerning the classical propositional case to the first-order level and, very briefly, to the case of non-classical logics.

Some, but not all, of the results presented in this paper were already made available in the research report [28].

2 The Logical Basis

We remain at the propositional level, and we consider the case of Classical Propositional Logic (hereafter: CPL). We assume that CPL is expressed in a language characterized as follows.

The vocabulary of the language comprises a countably infinite set Var of propositional variables, the connectives: $\neg, \vee, \wedge, \rightarrow$, and brackets. The set Form of *well-formed formulas* (wffs) of the language is the smallest set that includes Var and satisfies the following conditions: (1) if $A \in$ Form, then '$\neg A$' $\in$ Form; (2) if $A, B \in$ Form, then '$(A \otimes B)$' $\in$ Form, where $\otimes$ is any of the connectives: $\vee, \wedge, \rightarrow$. We adopt the usual conventions concerning omitting brackets. We use A, B, C, D, with subscripts when needed, as metalanguage variables for wffs, and X, Y, W, Z, with or without subscripts or superscripts, as metalanguage variables for sets of wffs. The letters p, q, r, s, t are exemplary elements of Var.

By a proper superset of a set of wffs X we mean a set of wffs Z such that X is a proper subset of Z. For the sake of brevity, we adopt the following notational conventions:

- we write X, Y instead of $X \cup Y$,
- X, A abbreviates $X \cup \{A\}$,
- $X_{\ominus A}$ abbreviates $X \setminus \{A\}$.

These conventions will be applied as long as there is no risk of a misunderstanding.

The inscriptions $\bigwedge X$ and $\bigvee Y$ refer to a conjunction of all the wffs in a non-empty and finite set of wffs X and to a disjunction of all the wffs in X, respectively. If X is a singleton set, $\{A\}$, then $\bigwedge X = \bigvee X = A$.

Let $\mathbf{1}$ stand for truth and $\mathbf{0}$ for falsity. A CPL-*valuation* is a function $v : \text{Form} \mapsto \{\mathbf{1}, \mathbf{0}\}$ satisfying the following conditions: (a) $v(\neg A) = \mathbf{1}$ iff $v(A) = \mathbf{0}$; (b) $v(A \vee B) = \mathbf{1}$ iff $v(A) = \mathbf{1}$ or $v(B) = \mathbf{1}$; (c) $v(A \wedge B) = \mathbf{1}$ iff $v(A) = \mathbf{1}$ and $v(B) = \mathbf{1}$; (d) $v(A \rightarrow B) = \mathbf{1}$ iff $v(A) = \mathbf{0}$ or $v(B) = \mathbf{1}$. Remark that the domain of v includes Var.

For brevity, in what follows we will be omitting references to CPL. Unless otherwise stated, the semantic relations analysed are supposed to hold between sets of CPL-wffs, or sets of CPL-wffs and single CPL-wffs. By valuations we will mean CPL-valuations.

We define:

Definition 1 (Sc-entailment). $X \models A$ *iff for each valuation* v:

- *if* $v(B) = \mathbf{1}$ *for every* $B \in X$, *then* $v(A) = \mathbf{1}$.

Wffs A and B are *logically equivalent* iff $A \models B$ and $B \models A$. Sets of wffs, X and Y, are logically equivalent iff they have exactly the same models, where a model of a set of wffs is a valuation which makes true all the wffs in the set.

Definition 2 (Mc-entailment). $X \models\mid Y$ *iff for each valuation* v:

- *if* $v(B) = \mathbf{1}$ *for every* $B \in X$, *then* $v(A) = \mathbf{1}$ *for at least one* $A \in Y$.

Definition 3 (Consistency, inconsistency, validity, and contingence). *A set of wffs* X *is* consistent *iff there exists a valuation* v *such that for each* $A \in X$, $v(A) = \mathbf{1}$; *otherwise* X *is* inconsistent. *A wff* B *is*:

1. consistent *iff the singleton set* $\{B\}$ *is consistent*,
2. inconsistent *iff the singleton set* $\{B\}$ *is inconsistent*,

3. valid *iff for each valuation* v, $v(B) = 1$,

4. contingent *iff B is neither inconsistent nor valid.*

Remark 1. Consistent wffs construed in the above manner are often called satisfiable wffs. The category of contingent wffs comprises wffs which are satisfiable, but not valid.

3 Strong Multiple-Conclusion Entailment

3.1 Definition and the Adequacy Issue

We use $\Vdash$ as the symbol for strong mc-entailment, and we define the relation as follows:[4]

Definition 4 (Strong mc-entailment). $X \Vdash Y$ *iff*

1. $X \models Y$, and

2. for each $A \in X : X_{\ominus A} \not\models Y$, and

3. for each $B \in Y : X \not\models Y_{\ominus B}$.

The consecutive clauses of the above definition express the following intuitions: the hypothetical truth of all the wffs in X warrants the existence of at least one true wff in Y, yet the warranty disappears as X decreases or Y decreases. In other words, X and Y are minimal sets under the warranty provided by the clause 1.

Here are simple examples:

$$\{p\} \Vdash \{p\} \qquad (1)$$
$$\{p, p \to q\} \Vdash \{q\} \qquad (2)$$
$$\{p \vee q, \neg p\} \Vdash \{q\} \qquad (3)$$
$$\emptyset \Vdash \{p, \neg p\} \qquad (4)$$
$$\{p, \neg p\} \Vdash \emptyset \qquad (5)$$
$$\emptyset \Vdash \{p \vee \neg p\} \qquad (6)$$
$$\{\neg p, \neg q, p \vee q\} \Vdash \emptyset \qquad (7)$$
$$\emptyset \Vdash \{p, q, \neg(p \vee q)\} \qquad (8)$$

---

[4] Recall that $X_{\ominus A}$ abbreviates $X \setminus \{A\}$, and similarly for Y.

$$\{p \vee q\} \mathrel{\Vert\!\prec} \{p, q\} \tag{9}$$
$$\{p \vee q\} \mathrel{\Vert\!\prec} \{p \wedge q, p \wedge \neg q, \neg p \wedge q\} \tag{10}$$
$$\{p \wedge q \to r, \neg r\} \mathrel{\Vert\!\prec} \{\neg p, \neg q\} \tag{11}$$
$$\{p \wedge (q \vee r)\} \mathrel{\Vert\!\prec} \{p \wedge q, p \wedge r\} \tag{12}$$
$$\{p \vee (q \vee r)\} \mathrel{\Vert\!\prec} \{p \vee q, p \vee r\} \tag{13}$$
$$\{\neg(p \wedge (q \wedge r))\} \mathrel{\Vert\!\prec} \{\neg p \vee \neg q, \neg p \vee \neg r\} \tag{14}$$
$$\{p \vee (q \vee r)\} \mathrel{\Vert\!\prec} \{(p \vee q) \wedge (p \vee r), q \vee r\} \tag{15}$$

Note that $\emptyset \mathrel{\Vert\!\not\prec} \emptyset$, as $\emptyset \mathrel{\not\models} \emptyset$.

Since the empty set has no proper subsets, and each proper subset of a non-empty set is included in a maximal proper subset of the set, it is clear that the following is true:

Corollary 1. $X \mathrel{\Vert\!\prec} Y$ iff $X \models Y$ and the following conditions hold:

1. there is no proper subset Z of X such that $Z \models Y$,

2. there is no proper subset W of Y such that $X \models W$.

Due to the monotonicity of "standard" mc-entailment, $\models$, we have:

Corollary 2. If $X \mathrel{\Vert\!\prec} Y$, then:

1. $Z \mathrel{\Vert\!\not\prec} Y$, where Z is either a proper subset or a proper superset of X,

2. $X \mathrel{\Vert\!\not\prec} W$, where W is either a proper subset or a proper superset of Y.

Thus strong mc-entailment, $\mathrel{\Vert\!\prec}$, is neither left-monotone nor right-monotone. The examples presented below witness this:

$$\{p, p \to q \vee r\} \mathrel{\Vert\!\prec} \{q, r\} \tag{16}$$
$$\{p, p \to q \vee r, \neg q\} \mathrel{\Vert\!\not\prec} \{q, r\} \tag{17}$$
$$\{p, p \to q \vee r\} \mathrel{\Vert\!\not\prec} \{q, r, q \vee r\} \tag{18}$$

Observe that the following are true:[5]

$$\{p, \neg p\} \mathrel{\Vert\!\not\prec} \{q\} \tag{19}$$
$$\{p\} \mathrel{\Vert\!\not\prec} \{p \vee \neg p\} \tag{20}$$

Thus it is neither the case that any inconsistent set of wffs strongly mc-entails any set of wffs nor it is the case that a set which contains a valid wff is strongly mc-entailed by any set of wffs. Hence strong mc-entailment is free of the drawbacks (I') and (II') pointed out in section 1.2.

---

[5] As for (19), $\{q\} \setminus \{q\} = \emptyset$, but we have $\{p, \neg p\} \models \emptyset$. In the case of (20) we have $\emptyset \models \{p \vee \neg p\}$.

3.1.1 Strong Mc-entailment, Perfect Validity, and Tennant's Entailments

Corollary 1 yields that our concept of strong mc-entailment is akin to (but not identical with) the concept of perfectly valid sequent introduced in [25], p. 185.

Assume for a moment that sequents are simply pairs of sets of wffs. A proper subsequent of a sequent $X : Y$ is a sequent resulting from it by removing at least one wff from X or from Y. Tennant's definition of validity of a sequent $X : Y$ amounts to the presence of mc-entailment of Y from X. A sequent $X : Y$ is *perfectly valid* iff $X : Y$ is valid and no proper subsequent of $X : Y$ is valid. Thus, by Corollary 1, a sequent $X : Y$ is perfectly valid iff $X \Vdash Y$ holds.[6]

However, perfect validity performs an auxiliary role in [25]. The central concept is that of sequent being an entailment. A sequent $X : Y$ is an *entailment* just in case $X : Y$ has a perfectly valid suprasequent. A sequent $Z : W$ is a suprasequent of the sequent $X : Y$ iff for some substitution s, $s(Z) = X$ and $s(W) = Y$. Tennant builds a sequent calculus which is sound and complete w.r.t. entailments construed in the above manner. A proof-theoretic account of perfectly valid sequents is also given by means of the so-called perfect proofs.

In this paper we will concentrate on a semantic analysis of strong mc-entailment or, if you prefer, perfect validity. A proof-theoretic account of strong mc-entailment, different from that offered by Tennant for perfect validity, will be also provided.

3.2 Basic Properties of Strong Mc-entailment

Let us first note:

Corollary 3. *Let A, B be logically equivalent wffs.*

1. *If $A \in X$ and $X \Vdash Y$, then $X_{\ominus A} \cup \{B\} \Vdash Y$.*

2. *If $A \in Y$ and $X \Vdash Y$, then $X \Vdash Y_{\ominus A} \cup \{B\}$.*

Thus logically equivalent wffs are replaceable in the context of strong mc-entailment. Needless to say, replaceability may fail for logical equivalence of sets of wffs. This

---

[6]But if the concept of proper subsequent is to be understood differently (i.e. $X' : Y'$ is a proper subsequent of $X : Y$ just in case $X' \subsetneq X$ or $Y' \subsetneq Y$, one needs the condition:

$$X' \cup Y' \subseteq X \cup Y$$

in order to pass from perfect validity to strong mc-entailment. Tennant does not provide an explicit definition of the notion of proper subsequent used.

is not surprising, as strong mc-entailment is a "hybrid" notion, defined in terms of semantic as well as set-theoretic clauses.[7]

Corollary 4. $\{A\} \Vvdash \{A\}$ iff A is contingent.

Proof. Clearly, $\{A\} \models \{A\}$, and $\{A\}_{\ominus A} = \emptyset$. On the other hand, A is not valid iff $\emptyset \not\models \{A\}$, and A is not inconsistent iff $\{A\} \not\models \emptyset$. □

However, the overlap/reflexivity condition is not satisfied in the case of non-singleton sets.

Corollary 5. If X has at least two elements, then $X \not\Vvdash X$.

Proof. Suppose otherwise. It follows that $X_{\ominus A} \not\models X$, where $A \in X$. But, as X has at least two elements, it holds that $X_{\ominus A} \cap X \neq \emptyset$ and hence $X_{\ominus A} \models X$. A contradiction. □

Let us now prove

Corollary 6. If $X \Vvdash Y$ and X is inconsistent, then $Y = \emptyset$.

Proof. Let $X \Vvdash Y$. Thus $X \models Y$. Assume that X is inconsistent. Suppose that $Y \neq \emptyset$. Thus $\emptyset$ is a proper subset of Y. However, $X \models \emptyset$ (since X is inconsistent) and hence $X \not\Vvdash Y$ due to Corollary 1. So $Y = \emptyset$. □

Thus an inconsistent set strongly mc-entails, if any, only the empty set. If any, since there are inconsistent sets that do not strongly mc-entail even the empty set. For instance, the set $\{p \wedge \neg p, p\}$ does not strongly entail the empty set because we still have $\{p \wedge \neg p\} \models \emptyset$. As we will see, only minimally inconsistent sets strongly mc-entail the empty set.

Remark 2. There exist strongly mc-entailed inconsistent sets of wffs. Examples (4) and (8) presented above support this claim. Here are examples which do not involve the empty set:

$$\{p\} \Vvdash \{p \wedge q, p \wedge \neg q\} \tag{21}$$

$$\{\neg(p \wedge q), p \vee q\} \Vvdash \{p \wedge \neg q, \neg p \wedge q\} \tag{22}$$

---

[7]Such a solution has obvious vices, but also some virtues; see sections 4.2.3 and 5.1 below.

3.2.1 Contingent, Valid, and Inconsistent Wffs

Interestingly enough, strong mc-entailment between non-empty sets of wffs involves sets which comprise contingent wffs only. The following holds:

Theorem 1 (Contingency). *Let $X \Vdash Y$. If $X \neq \emptyset$ and $Y \neq \emptyset$, then each wff in $X \cup Y$ is contingent.*

Proof. Assume that $X \Vdash Y$, where X and Y are non-empty sets.

Suppose that X contains a valid wff, say, A. It follows that $X_{\ominus A} \models Y$ and therefore $X \nVdash Y$. Now suppose that X contains an inconsistent wff. Hence X is an inconsistent set. But $Y \neq \emptyset$. Thus, by Corollary 6, $X \nVdash Y$, which contradicts the assumption.

Therefore X contains contingent wffs only.

Suppose that a valid wff, say, A, belongs to Y. By assumption, $X \neq \emptyset$, so $\emptyset$ is a proper subset of X. Suppose that $Y = \{A\}$. Clearly, $\emptyset \models \{A\}$ due to the validity of A. Hence $X \nVdash \{A\}$. Now suppose that $Y \neq \{A\}$. As $Y \neq \emptyset$, it follows that $\{A\}$ is a proper subset of Y which, however, is mc-entailed by X since A is valid. Thus $X \nVdash Y$. Therefore no valid wff belongs to Y.

Finally, suppose that an inconsistent wff, B, belongs to Y. In this case $X \models Y$ yields $X \models Y_{\ominus B}$. As Y is, by assumption, non-empty, $Y_{\ominus B}$ is a proper subset of Y. It follows that $X \nVdash Y$. We arrive at a contradiction. Thus no wff in Y is inconsistent.

Therefore $X \cup Y$ contains contingent wffs only. □

What if either X or Y is empty? The answer is provided by:

Corollary 7.

1. *If $\emptyset \Vdash Y$, then either Y is a singleton set containing a valid wff, or Y is a non-singleton set comprising only contingent wffs.*

2. *If $X \Vdash \emptyset$, then either X is a singleton set containing an inconsistent wff, or X is a non-singleton set comprising only contingent wffs.*

Proof. If $\emptyset \Vdash Y$, then $Y \neq \emptyset$. Assume that Y is a singleton set, $\{C\}$. Since $\emptyset \Vdash \{C\}$ presupposes $\emptyset \models \{C\}$, it follows that C is a valid wff. Assume that Y is a non-singleton set. Suppose that Y contains a non-contingent wff, say, B. If B is valid, then $\emptyset \models \{B\}$ and hence $\emptyset \nVdash Y$. The situation is analogous when B is an inconsistent wff; in this case we would have $\emptyset \models Y_{\ominus B}$.

If $X \Vdash \emptyset$, then $X \neq \emptyset$. Assume that X is a singleton set, $\{C\}$. Thus $\{C\} \models \emptyset$ and hence C is an inconsistent wff. Assume that X is a non-singleton set. Suppose that X contains a non-contingent wff, say, A. Clearly, $X_{\ominus A}$ is a proper subset of X and so is $\{A\}$. Assume that A is valid. Thus $X_{\ominus A} \models \emptyset$ and hence $X \not\Vdash \emptyset$ does not hold. Now assume that A is inconsistent. Thus $\{A\} \models \emptyset$ and hence, again, $X \not\Vdash \emptyset$ is not the case. Therefore each wff in X is contingent provided that X is a non-singleton set. □

As for strong mc-entailment, non-contingent wffs come into play in two exceptional situations only.

Theorem 2. *Let* $X \Vdash Y$.

1. *If C is valid, then:* $C \in X \cup Y$ *iff* $X = \emptyset$ *and* $Y = \{C\}$.

2. *If C is inconsistent, then:* $C \in X \cup Y$ *iff* $X = \{C\}$ *and* $Y = \emptyset$.

Proof. Let C be a valid wff. Assume that $X \Vdash Y$ and $C \in X \cup Y$.

Suppose that $C \in X$. Hence $X \neq \emptyset$ and $X_{\ominus C}$ is a proper subset of X. If C is valid, then whatever is mc-entailed by X is also mc-entailed by $X_{\ominus C}$. So $X \not\Vdash Y$. We arrive at a contradiction. Therefore $C \notin X$ and thus $C \in Y$.

Suppose that $X \neq \emptyset$. Thus $\emptyset$ is a proper subset of X. Since C is valid and $C \in Y$, we have $\emptyset \models Y$. It follows that $X \not\Vdash Y$, contrary to the assumption. Therefore $X = \emptyset$. As $C \in Y$, it follows that Y does not comprise contingent wffs only. Hence $Y = \{C\}$ due to Corollary 7.

Needless to say, if $Y = \{C\}$, then $C \in X \cup Y$.

The proof of (*2*) goes along similar lines. □

According to Theorem 2, valid wffs can occur as elements of strongly mc-entailed sets, but these sets are always singleton sets which, moreover, are strongly mc-entailed only by the empty set. Similarly, if an inconsistent wff belongs to a strongly mc-entailing set, it is the only element of this set and the respective strongly mc-entailed set is empty. Moreover, valid wffs never occur in strongly mc-entailing sets, and inconsistent wffs never occur in strongly mc-entailed sets.

3.2.2 Strict Finiteness and Variable Sharing

We are dealing here with **CPL**, in which mc-entailment has the following properties:

(lf) *If* $X \models Y$, *then* $X_1 \models Y$ *for some finite subset* X_1 *of* X.

(rf) *If* $X \models Y$, *then* $X \models Y_1$ *for some finite subset* Y_1 *of* Y.

As for **CPL** (and other logics in which mc-entailment fulfils the above conditions), strong mc-entailment is strictly finitistic in the sense explained by:

Theorem 3 (Strict finiteness). *If $X \Vdash Y$, then X and Y are finite sets.*

Proof. Let $X \Vdash Y$.

Suppose that X is an infinite set. By Corollary 1, it follows that there is no finite subset of X which mc-entails Y. Hence $X \nVdash Y$ due to condition (If). But $X \Vdash Y$ yields $X \vDash Y$. So X is a finite set. Now suppose that Y is an infinite set. Hence, by Corollary 1, no finite subset of Y is mc-entailed by X. Thus $X \nVdash Y$ due to condition (rf). It follows that $X \Vdash Y$ does not hold, contrary to the assumption. So Y is a finite set as well. □

Our next theorem is strongly dependent on the fact that we consider here propositional formulas.

Notation. By $\text{Var}(A)$ we designate the set of all the propositional variables that occur in a wff A. $\text{Var}(X)$ designates the set of all the propositional variables that occur in the wffs which belong to a set of wffs X.

Theorem 4 (Variable sharing). *Let $X \Vdash Y$. If X and Y are non-empty sets, then $\text{Var}(X) \cap \text{Var}(Y) \neq \emptyset$.*

Proof. Let $X \Vdash Y$, where $X \neq \emptyset$ and $Y \neq \emptyset$.

If $X \neq \emptyset$, then, by Corollary 2, $\emptyset \nvDash Y$. By assumption, $Y \neq \emptyset$. So there exists a valuation, say, v^*, such that $v^*(B) = \mathbf{0}$ for any $B \in Y$. By Corollary 6, X is consistent. Hence there exists a valuation v such that $v(A) = \mathbf{1}$ for every $A \in X$.

Suppose that $\text{Var}(X) \cap \text{Var}(Y) = \emptyset$. Let v^+ be a valuation such that: (a) $v^+(p_i) = v^*(p_i)$ if $p_i \in \text{Var}(Y)$, (b) otherwise $v^+(p_i) = v(p_i)$. As $\text{Var}(X) \cap \text{Var}(Y) = \emptyset$, we have $v^+(A) = \mathbf{1}$ for every $A \in X$. On the other hand, $v^+(B) = \mathbf{0}$ for each $B \in Y$. Hence $X \nvDash Y$ and therefore $X \nVdash Y$. We arrive at a contradiction. □

So when strong mc-entailment between X and Y holds, the wffs in X share propositional variable(s) with the wffs in Y. However, Theorem 4 cannot be strengthened to the effect that $\text{Var}(Y) \subseteq \text{Var}(X)$ would be the case. Similarly, $\text{Var}(X) \subseteq \text{Var}(Y)$ does not generally hold.[8]

3.2.3 Partial Reduction to Minimally Inconsistent Sets

As long as a logic operating with the classical negation is concerned, there exist simple links between strong mc-entailment and minimally inconsistent sets:[9]

---

[8]For instance, we have $\{p \vee q\} \Vdash \{p, r \to q\}$ as well as $\{p \wedge q\} \Vdash \{p\}$.

[9]The concept of minimally inconsistent set has found natural applications is many areas, from philosophy of science (cf., e.g., [9]) to theoretical computer science, AI, and logic (see, e.g., [12],

Definition 5 (Minimally inconsistent set; MI-set). *A set of wffs X is minimally inconsistent iff X is inconsistent, but each proper subset of X is consistent.*

For brevity, we will be referring to minimally inconsistent sets as to MI-*sets*.

Note that $\emptyset$ is not a MI-set. Singleton MI-sets have inconsistent wffs as the (only) elements. Here are examples of non-singleton MI-sets:

$$\{p, \neg p\} \tag{23}$$
$$\{p \vee q, \neg p, \neg q\} \tag{24}$$
$$\{p \to q, p, \neg q\} \tag{25}$$
$$\{p \to q \vee r, p, \neg q, \neg r\} \tag{26}$$
$$\{p \to q, q \to r, \neg(p \to r)\} \tag{27}$$

Clearly, the following holds:

Corollary 8. *X is a MI-set iff X is inconsistent and for each $A \in X$, the set $X_{\ominus A}$ is consistent.*

Remark 3. As for CPL, any MI-set is finite. This is due to the fact that the following *compactness claim* holds for CPL:

(♣) *for each set of wffs Z: the set Z is consistent iff each finite subset of Z is consistent.*

However, there are logics for which the analogues of (♣) do not hold and thus finiteness is not a property of MI-sets in general.[10]

Notation. For brevity, we put:

$$\neg Y =_{df} \{\neg A : A \in Y\}$$

In the case of CPL, strong mc-entailment and MI-sets are linked in the following way:

Theorem 5. *$X \Vdash Y$ iff $X \cap \neg Y = \emptyset$ and $X, \neg Y$ is a MI-set.*

---

[3], [16]). Minimally inconsistent sets are also called *minimal unsatisfiable (sub)sets* or *unsatisfiable cores*.

[10]For example, in a logic that validates the ω-rule, a set of the form $\{\exists x Px\} \cup \{\neg Pa : a \in \mathsf{T}\}$, where P is a predicate and T is a (countably infinite) set of all closed terms of the language, is an infinite MI-set.

Proof. ($\Rightarrow$) Let $X \Vdash Y$. Suppose that $X \cap \neg Y \neq \emptyset$. Let $A \in X \cap \neg Y$. Thus $A = \neg B$ for some $B \in Y$. Let $Y^* = Y_{\ominus B}$. From $X \Vdash Y$ we get $X \models Y^*, B$. Therefore $X, \neg B \models Y^*$, that is, $X, A \models Y^*$. But $X, A = X$, since $A \in X$. Hence X mc-entails the proper subset Y^* of Y. It follows that $X \nVdash Y$. We arrive at a contradiction. Therefore $X \cap \neg Y = \emptyset$.

If $X \Vdash Y$, then $X \models Y$ and thus the set $X, \neg Y$ is inconsistent. Let us designate the set $X, \neg Y$ by Z.

If $A \in Z$, then $A \in X$ or $A \in \neg Y$.

Assume that $A \in X$. By the clause 2 of Definition 4, $X_{\ominus A} \nmodels Y$ and thus the set $X_{\ominus A}, \neg Y$ is consistent, that is, $Z_{\ominus A}$ is consistent.

Now assume that $A \in \neg Y$. Hence $A = \neg B$ for some $B \in Y$. By the clause 3 of Definition 4, $X \nmodels Y_{\ominus B}$. Thus the set $X, \neg(Y_{\ominus B})$ is consistent. Yet, $X, \neg(Y_{\ominus B}) = Z_{\ominus A}$. Hence the set $Z_{\ominus A}$ is consistent.

By Corollary 8, $X, \neg Y$ is thus a MI-set.

($\Leftarrow$) Assume that $X \cap \neg Y = \emptyset$ and $X, \neg Y$ is a MI-set. From the latter it follows that $X \models Y$.

Again, let $Z = X, \neg Y$.

Suppose that $X_{\ominus A} \models Y$ for some $A \in X$. Then the set $X_{\ominus A}, \neg Y$ is inconsistent. Yet, since $X \cap \neg Y = \emptyset$, the set $X_{\ominus A}, \neg Y$ is a proper subset of Z. Thus Z is not a MI-set. A contradiction.

Now suppose that $X \models Y_{\ominus B}$ for some $B \in Y$. Let us designate $Y_{\ominus B}$ by Y^*. As $X \models Y^*$ holds, the set $X, \neg Y^*$ is inconsistent. But $X \cap \neg Y = \emptyset$, so $\neg B$ does not belong to X. Hence the set $X, \neg Y^*$ is a proper subset of Z. Thus Z is not a MI-set. A contradiction again.

Therefore $X \Vdash Y$. □

Theorem 5 yields:

Corollary 9.

1. $X \Vdash \emptyset$ iff X is a MI-set.

2. $\emptyset \Vdash Y$ iff $\neg Y$ is a MI-set.

Remark 4. As the second part of the proof of Theorem 5 shows, one can get $X \Vdash Y$ from the fact that $X, \neg Y$ is a MI-set *on the condition* that $X \cap \neg Y = \emptyset$ holds. This condition is a necessary one. For example, let $X = \{p \vee q, \neg p, \neg q\}$ and $Y = \{p, q\}$. Then $\neg Y = \{\neg p, \neg q\}$ and hence $X, \neg Y = X$. As X is a MI-set, so is $X, \neg Y$. However, $X \nVdash Y$, since $\{p \vee q\} \models \{p, q\}$. On the other hand, $X \cap \neg Y = \{\neg p, \neg q\} \neq \emptyset$.

There exist MI-sets which do not contain wffs beginning with negation, i.e. wffs of the form $\neg B$. Here are simple examples:

$$\{p \to q, p \land \neg q\}$$
$$\{p, p \to q, p \to \neg q\}$$

It may seem that such MI are "useless" in showing that strong mc-entailment holds. But this is wrong. The corollary below explains why.

Corollary 10. *If X, Y is a MI-set and $X \cap Y = \emptyset$, then $X \parallel\!\prec \neg Y$.*

Proof. Clearly, if X, Y is a MI-set, then $X, \neg(\neg Y)$ is a MI-set. Suppose that $X \cap \neg(\neg Y) \neq \emptyset$. So there exists $A \in X$ such that $A = \neg\neg B$ for some $B \in Y$, and $A \in \neg(\neg Y)$. As X, Y is a MI-set and $X \cap Y = \emptyset$, we have $B \notin X$ and thus the set $X, Y_{\ominus B}$ is consistent. Hence $X, (\neg(\neg Y))_{\ominus \neg\neg B}$ is a consistent set as well. But $X, (\neg(\neg Y))_{\ominus \neg\neg B} = X, \neg(\neg Y)$, since $A = \neg\neg B$ and $A \in X$. It follows that $X, \neg(\neg Y)$ is not a MI-set. We arrive at a contradiction. Thus $X \cap \neg(\neg Y) = \emptyset$. As $X, \neg(\neg Y)$ is a MI-set, by Theorem 5 we get $X \parallel\!\prec \neg Y$. □

3.2.4 Independence and Deduction

Observe that if X strongly mc-entails Y, then neither X nor Y contains syntactically distinct wffs which are logically equivalent, i.e. entail each other. The reason is that a MI-set never includes logically equivalent wffs. We can also prove more:

Theorem 6 (Independence). *Let $X \parallel\!\prec Y$, and let A, B be syntactically distinct wffs.*

1. *If $A, B \in X$ and $Y \neq \emptyset$, then $A \not\models B$ and $A \not\models \neg B$.*

2. *If $A, B \in Y$, then $A \not\models B$, and $\neg A \not\models B$ provided that $\{A, B\} \neq Y$.*

Proof. If $X \parallel\!\prec Y$, then, by Theorem 5, $X, \neg Y$ is a MI-set and $X \cap \neg Y = \emptyset$.

Let $A, B \in X$. Thus the set $X_{\ominus B}, \neg Y$ is consistent and, due to the fact that $X, \neg Y$ is inconsistent, $X_{\ominus B}, \neg Y \models \neg B$. But $A \in X_{\ominus B}$. Therefore $X_{\ominus B}, \neg Y \models A$. Hence $A \not\models B$.

As $B \in X$, we have $X \models B$. Suppose that $A \models \neg B$. Since $A \in X$, it follows that $X \models \neg B$. Thus X is an inconsistent set and, as $Y \neq \emptyset$, we get $X \parallel\!\not\prec Y$.

Let $A, B \in Y$. It follows that $\neg A, \neg B \in \neg Y$. By Theorem 5, $X, \neg Y$ is a MI-set and hence an inconsistent set. Thus $X, \neg(Y_{\ominus A}) \models A$. However, the set $X, \neg(Y_{\ominus A})$, as a proper subset of the MI-set in question, is consistent. On the other hand, '$\neg B$' $\in \neg(Y_{\ominus A})$. It follows that $X, \neg(Y_{\ominus A}) \models \neg B$. Therefore $A \not\models B$.

Assume that $\{A, B\} \neq Y$. Suppose that $\neg A \models B$. It follows that $\emptyset \models \neg A \to B$ and hence $\emptyset \models \{A, B\}$. As $\{A, B\} \neq Y$, we get $X \parallel\!\not\prec Y$. □

Notation. For conciseness, let us introduce the following notational convention:

$$\lceil A \to W \rceil =_{df} \begin{cases} \{\neg A\} & \text{if } W = \emptyset, \\ \{A \to B : B \in W\} & \text{if } W \neq \emptyset. \end{cases}$$

One can easily show that the following holds:

Corollary 11. $Z, A \models W$ iff $Z \models \lceil A \to W \rceil$.

As a consequence we get:

Theorem 7 (Deduction for strong mc-entailment). *Let $A \notin X$. If $X, A \Vvdash Y$, then $X \Vvdash \lceil A \to Y \rceil$.*

Proof. Assume that $X, A \Vvdash Y$. If $X, A \models Y$, then, by Corollary 11, $X \models \lceil A \to Y \rceil$. Let $B \in X$. Since, by assumption, $A \notin X$, it follows that $A \neq B$. Thus $X_{\ominus B}, A$ is a proper subset of X, A. As $X, A \Vvdash Y$ holds, we have $X_{\ominus B}, A \not\models Y$. Hence, by Corollary 11 again, $X_{\ominus B} \not\models \lceil A \to Y \rceil$. Let $C \in Y$. Thus $X, A \not\models Y_{\ominus C}$. Therefore, by Corollary 11, $X \not\models \lceil A \to Y_{\ominus C} \rceil$. Hence $X \Vvdash \lceil A \to Y \rceil$. □

Note that the converse of Theorem 7 is not true. For example, $\emptyset \Vvdash \{p \to q, p \to \neg q\}$ holds, but $\{p\} \Vvdash \{q, \neg q\}$ is not the case. However, the following is true:

Corollary 12. *If $X \Vvdash \lceil A \to Y \rceil$ and $X \not\models \neg A$ as well as $X \not\models Y$, then $X, A \Vvdash Y$.*

Proof. Suppose that $Y = \emptyset$. Hence $X \Vvdash \{\neg A\}$. Thus $X \models \neg A$. But, by assumption, $X \not\models \neg A$. So $Y \neq \emptyset$.

If $X \Vvdash \lceil A \to Y \rceil$, then, by Definition 4 and Corollary 11, $X, A \models Y$ and $X_{\ominus B} \cup \{A\} \not\models Y$ for any $B \in X$. By assumption, $X \not\models Y$. It follows that for every $C \in X, A$ we have $X, A \setminus \{C\} \not\models Y$. Now suppose that $X, A \models Y_{\ominus D}$ is the case for some $D \in Y$. There are two possibilities: (a) $Y_{\ominus D} = \emptyset$ and (b) $Y_{\ominus D} \neq \emptyset$. Assume that (a) holds. It follows that the set X, A is inconsistent. But, by assumption, $X \not\models \neg A$ and hence the set X, A is consistent. So (a) does not hold. It follows that Y is not a singleton set. Assume that (b) is the case. Therefore, by Corollary 11, $X \models \lceil A \to Y_{\ominus D} \rceil$. As $\lceil A \to Y_{\ominus D} \rceil$ is a proper subset of $\lceil A \to Y \rceil$, it follows that $X \not\Vvdash \lceil A \to Y \rceil$. So we arrive at a contradiction again. Hence $X, A \not\models Y_{\ominus D}$ for every $D \in Y$. As all the clauses of Definition 4 are fulfilled w.r.t. X, A and Y, we conclude that $X, A \Vvdash Y$ holds. □

As the proof of Corollary 12 shows, the assumption "$X \not\models \neg A$" is dispensable when Y is neither a singleton set nor the empty set.

Corollary 13. *Let Y be a finite and at least two-element set of wffs. If $X \Vdash \lceil A \to Y \rceil$ and $X \not\Vdash Y$, then $X, A \Vdash Y$.*

Finally, observe that $\Vdash$ is not closed under uniform substitution. A simple example illustrates this. Clearly, $\{p\} \Vdash \{p\}$ is the case. But $\{p \wedge \neg p\} \Vdash \{p \wedge \neg p\}$ does not hold (cf. Corollary 6). Needless to say, $p \wedge \neg p$ results from p by substitution.

4 Strong Single-Conclusion Entailment

4.1 Definition and the Adequacy Issue

Sc-entailment traditionally construed can be identified with mc-entailment of a singleton set. Similarly, it seems natural to define strong sc-entailment as strong mc-entailment of a singleton set.

We use $\vdash$ as the symbol for strong sc-entailment.

Definition 6 (Strong sc-entailment). *$X \vdash B$ iff $X \Vdash \{B\}$.*

For brevity, we will write $A \vdash B$ instead of $\{A\} \vdash B$.

As an immediate consequence of Definition 6 and Theorem 5 one gets:

Theorem 8. *$X \vdash B$ iff '$\neg B$' $\notin X$ and $X, \neg B$ is a MI-set.*

Note that the transition from right to left requires '$\neg B$' $\notin X$ to hold. For example, although

$$\{p, p \to \neg q, \neg\neg q\} \cup \{\neg\neg q\} \tag{28}$$

is a MI-set, $\{p, p \to \neg q, \neg\neg q\} \vdash \neg q$ does not hold, since '$\neg\neg q$' $\in \{p, p \to \neg q, \neg\neg q\}$.

The following is true:

Corollary 14. *$X \vdash B$ iff*

1. *$X \models B$ and*

2. *for each proper subset Z of X: $Z \not\models B$, and*

3. *X is consistent.*

Proof. Clearly, $X \models B$ holds iff $X \models \{B\}$ is the case.

Clause 2 holds due to Corollary 1. On the other hand, clause 2 yields that there is no $A \in X$ such that $X_{\ominus A} \models \{B\}$.

Since $\{B\} \setminus \{B\} = \emptyset$, clause 3 of Definition 4 and clause 3 of the above corollary are equivalent for $Y = \{B\}$. □

Strong sc-entailment is not monotone. As a matter of fact, it is "antimonotone" in a sense explained by:

Corollary 15. *If $X \mid\kern-.4em\sim B$ and $X \subseteq Y$, where $Y \neq X$, then $Y \not\mid\kern-.4em\sim B$.*

Proof. By Definition 6 and Corollary 2. □

As we pointed out in section 1.1, the monotonicity of entailment contravenes, in a sense, the semantic entrenchment idea, since it allows semantically irrelevant wffs to occur among premises. In the case of strong sc-entailment, however, the difficulty is solved in a radical way: a strongly sc-entailing set is "minimal" with regard to the transmission of truth and, since no proper superset of a set X that strongly sc-entails a wff B strongly sc-entails B as well, adding an "irrelevant" wff to X results in the lack of strong sc-entailment of B from X enriched in this way.

By the clause 2 of Corollary 14, each proper subset of a strongly sc-entailing set is consistent. Strong sc- and mc-entailment do not differ in this respect. As we have seen, however, there exist strongly mc-entailing sets which are inconsistent (each of them strongly mc-entails only the empty set, however). According to the clause 3 of Corollary 14, this never happens in the case of strong sc-entailment. Anyway, strong sc-entailment is free of the drawback (I) pointed out in section 1.1. Let us add: free, again, in a radical way, since inconsistent sets do not strongly sc-entail any wffs. As an immediate consequence of Corollary 6 one gets:

Corollary 16. *No wff is strongly sc-entailed by an inconsistent set of wffs.*

Thus no inconsistent wff belongs to a sc-entailing set, and a singleton set which comprises an inconsistent wff does not strongly sc-entail any wff. In particular, neither $A \wedge \neg A \mid\kern-.4em\sim A$ nor $\{A, \neg A\} \mid\kern-.4em\sim A$ holds, regardless of what A is. Similarly, there is no B such that $A \wedge \neg A \mid\kern-.4em\sim B$ or $\{A, \neg A\} \mid\kern-.4em\sim B$.

Observe that the following holds as well:

Corollary 17. *There is no set of wffs that strongly sc-entails an inconsistent wff.*

Proof. By Definition 6 and Theorem 2. □

Thus inconsistencies are outside the realm of strong sc-entailment: no inconsistent set belongs to the domain of $\mid\kern-.4em\sim$ and no inconsistent wff belongs to the range of the relation. No doubt, a paraconsistent logician will dislike strong sc-entailment.

The case of validities is slightly more complicated. By Theorem 2 we get:

Corollary 18. *If $X \mid\kern-.4em\sim B$, then no wff in X is valid.*

Corollary 19. *If B is valid and $X \mid\kern-.4em\sim B$, then $X = \emptyset$.*

One can prove that valid wffs are exactly these wffs which are strongly sc-entainled only by the empty set.

Corollary 20. *A wff B is valid iff $\emptyset \mathrel{\mid\!\!\sim} B$ and $X \mathrel{\mid\!\!\not\sim} B$ for any $X \neq \emptyset$.*

Proof. Let B be a valid wff. Thus $\{\neg B\}$ is a Ml-set, and hence, by Corollary 9, $\emptyset \mathrel{\mid\!\!\sim} B$. Thus $X \mathrel{\mid\!\!\not\sim} B$ for any $X \neq \emptyset$. On the other hand, if $\emptyset \mathrel{\mid\!\!\sim} B$, then $\emptyset \models B$ and hence B is valid. $\square$

As for valid wffs, Corollary 19 yields that the difference between strong sc-entailment and sc-entailment simpliciter lies in the fact that valid wffs are strongly sc-entailed *only* by the empty set. Thus, in particular, valid wffs are not strongly sc-entailed by sets of valid wffs. Moreover, a valid wff is not sc-entailed by any set of wffs to which a valid wff belongs to.

4.2 Some Properties of Strong Sc-entailment

Since strong sc-entailment is defined in terms of strong mc-entailment, one can easily derive the following corollaries from the corresponding results presented in sections 3.2.1, 3.2.2, and 3.2.4.

Corollary 21. *Let A, B be logically equivalent wffs.*

1. *If $A \in X$ and $X \mathrel{\mid\!\!\sim} C$, then $X_{\ominus A} \cup \{B\} \mathrel{\mid\!\!\sim} C$.*

2. *If $X \mathrel{\mid\!\!\sim} A$, then $X \mathrel{\mid\!\!\sim} B$.*

Corollary 22 (Contingency for $\mathrel{\mid\!\!\sim}$)**.** *If $X \mathrel{\mid\!\!\sim} B$ and $X \neq \emptyset$, then each wff in $X \cup \{B\}$ is contingent.*

Corollary 23 (Strict finiteness of $\mathrel{\mid\!\!\sim}$)**.** *If $X \mathrel{\mid\!\!\sim} B$, then X is a finite set.*

Corollary 24 (Variable sharing for $\mathrel{\mid\!\!\sim}$)**.** *If $X \mathrel{\mid\!\!\sim} B$ and $X \neq \emptyset$, then $\mathtt{Var}(X) \cap \mathtt{Var}(B) \neq \emptyset$.*

Corollary 25 (Independence for $\mathrel{\mid\!\!\sim}$)**.** *Let $X \mathrel{\mid\!\!\sim} B$. If A, C are syntactically distinct wffs that belong to X, then $A \not\models C$ and $A \not\models \neg C$.*

Corollary 26 (Deduction for strong sc-entailment)**.** *Let $A \notin X$. If $X, A \mathrel{\mid\!\!\sim} B$, then $X \mathrel{\mid\!\!\sim} A \to B$.*

The converse of Corollary 26 is not true. For instance, $\emptyset \mathrel{\mid\!\!\sim} p \wedge \neg p \to q$ holds, but $p \wedge \neg p \mathrel{\mid\!\!\sim} q$ does not hold. Yet, there are cases in which $X \mathrel{\mid\!\!\sim} A \to B$ yields $X, A \mathrel{\mid\!\!\sim} B$. Corollary 12 implies:

Corollary 27. *If $X \mathrel{\mid\kern-0.4em\sim} A \to B$, and $X \not\models \neg A$ as well as $X \not\models B$, then $X, A \mathrel{\mid\kern-0.4em\sim} B$.*

Thus we get:

Corollary 28. *Let $A \notin X$, and $X \not\models \neg A$ as well as $X \not\models B$. Then $X, A \mathrel{\mid\kern-0.4em\sim} B$ iff $X \mathrel{\mid\kern-0.4em\sim} A \to B$.*

Proof. By corollaries 26 and 27. □

Strong sc-entailment is, in a sense, closed under detachment.

Corollary 29 (Detachment for $\mathrel{\mid\kern-0.4em\sim}$). *If $X \mathrel{\mid\kern-0.4em\sim} A \to B$ and $X \mathrel{\mid\kern-0.4em\sim} A$, then $X \mathrel{\mid\kern-0.4em\sim} B$.*

Proof. Either $X \mathrel{\mid\kern-0.4em\sim} A \to B$ or $X \mathrel{\mid\kern-0.4em\sim} A$ warrants the consistency of X, and together they yield that $X \models B$ holds.

Assume that $X \neq \emptyset$. Let C be an arbitrary but fixed element of X. From $X \mathrel{\mid\kern-0.4em\sim} A \to B$ we get $X_{\odot C} \not\models A \to B$. It follows that $X_{\odot C} \not\models B$. Thus $X \mathrel{\mid\kern-0.4em\sim} B$.

Now assume that $X = \emptyset$. In this case B is a valid wff. Therefore $\emptyset \mathrel{\mid\kern-0.4em\sim} B$ due to Corollary 20, that is, $X \mathrel{\mid\kern-0.4em\sim} B$. □

Observe that one can also prove that $X \mathrel{\mid\kern-0.4em\sim} A \to B$ and $X \models A$ yield $X \mathrel{\mid\kern-0.4em\sim} B$.

4.2.1 Strong Sc-entailment from Singleton Sets

Strong sc-entailment from single wffs (more precisely, from singleton sets of wffs) has some properties which strong sc-entailment from non-singleton sets lack.

Corollary 30. *The following are equivalent:*

1. *$A \mathrel{\mid\kern-0.4em\sim} B$,*

2. *$A \models B$ and A, B are contingent wffs.*

Proof. The implication from (1) to (2) is due to Definition 6 and Corollary 22. As for the passage from (2) to (1), it suffices to observe that the contingency of A warrants the consistency of $\{A\}$, while the contingency of B guarantees that $\emptyset \Vdash \{B\}$ does not hold. □

One cannot generalize Corollary 30 to non-singleton sets. The contingency of all the wffs belonging to a (non-empty) non-singleton set of wffs X warrants neither the consistency of X itself nor the lack of entailment of B from proper subset(s) of X.

Coming back to sc-entailment from single wffs. The lack of strong sc-entailment in the presence of standard sc-entailment tells us more about the wffs involved than Corollary 30 does.

Corollary 31. *If $A \models B$, but $A \mathrel{\not\mathrel{\mid\kern-3pt\sim}} B$, then A is inconsistent or B is valid.*

Proof. If $A \models B$ and $A \mathrel{\not\mathrel{\mid\kern-3pt\sim}} B$, then $\{A\}$ is an inconsistent set or $\emptyset \models B$. So A is inconsistent or B is valid. □

When X is a non-empty set having more that one element, the lack of $X \mathrel{\mid\kern-3pt\sim} B$ in the presence of $X \models B$ implies that X is inconsistent or B is entailed by some proper subset of X.

Finally, let us notice the following:

Corollary 32. *If $A \mathrel{\mid\kern-3pt\sim} B$ and $B \mathrel{\mid\kern-3pt\sim} C$, then $A \mathrel{\mid\kern-3pt\sim} C$.*

Proof. Certainly, $A \models B$ and $B \models C$ yields $A \models C$. By Corollary 30, $A \mathrel{\mid\kern-3pt\sim} B$ warrants the contingency of A, while $B \mathrel{\mid\kern-3pt\sim} C$ yields the contingency of C. So $A \mathrel{\mid\kern-3pt\sim} C$ due to Corollary 30. □

Observe that when X has more than one element, the passage from $X \mathrel{\mid\kern-3pt\sim} B$ and $B \mathrel{\mid\kern-3pt\sim} C$ to $X \mathrel{\mid\kern-3pt\sim} C$ requires an additional condition to be met, namely it must be ensured that for each $D \in X$, the set $X_{\ominus D}$ does not entail C.

4.2.2 Mutuality

As for CPL, mc-entailment of a non-empty finite set reduces to sc-entailment of a disjunction of all the elements of the set, i.e. if Y is a finite set and $Y \neq \emptyset$, then $X \models Y$ iff $X \models \bigvee Y$. But strong mc-entailment and strong sc-entailment are not linked in this way. For instance, we have:

$$p \mathrel{\mid\kern-3pt\sim} p \vee q \tag{29}$$

but we *do not* have:[11]

$$p \mathrel{\|\kern-3pt\sim} \{p, q\} \tag{30}$$

Strong mc- and sc-entailments are mutually linked in a quite different way, as the following theorem shows.

Theorem 9 (Mutuality).

1. *If $X \mathrel{\|\kern-3pt\sim} Y, B$, where $B \notin Y$, then $X, \neg Y \mathrel{\mid\kern-3pt\sim} B$.*

2. *If $X, \neg Y \mathrel{\mid\kern-3pt\sim} B$ and $X \cap \neg Y = \emptyset$, then $X \mathrel{\|\kern-3pt\sim} Y, B$.*

---

[11] (30) does not hold because $\{p\} \models (\{p, q\} \setminus \{q\})$.

Proof. If $X \mathrel{\|\!\!\sim} Y, B$, then, by Theorem 5, $X \cup (\neg Y \cup \{\neg B\})$ is a MI-set and $X \cap (\neg Y \cup \{\neg B\}) = \emptyset$. It follows that $(X \cup \neg Y) \cup \{\neg B\}$ is a MI-set and '$\neg B$' $\notin X$. By assumption, $B \notin Y$. So '$\neg B$' $\notin \neg Y$. Hence '$\neg B$' $\notin X, \neg Y$. Thus $X, \neg Y \mathrel{\triangleright\!\sim} B$ by Theorem 8.

If $X, \neg Y \mathrel{\triangleright\!\sim} B$, then, by Theorem 8, $(X \cup \neg Y) \cup \{\neg B\}$ is a MI-set and '$\neg B$' $\notin X \cup \neg Y$. Suppose that $X \cap (\neg Y \cup \{\neg B\}) \neq \emptyset$. As '$\neg B$' $\notin X \cup \neg Y$, it follows that $(X \cap \neg Y) \neq \emptyset$. On the other hand, by assumption $(X \cap \neg Y) = \emptyset$. Therefore $X \mathrel{\|\!\!\sim} Y, B$ due to Theorem 5. □

4.2.3 Conjunction vs. Set of Conjuncts

As for the standard sc-entailment based on Classical Logic, there is no scope difference between being entailed by a finite set of wffs and being entailed by a conjunction of all the wffs of this set. Although conjunction, $\wedge$, is semantically construed here in the classical manner (cf. Section 2), it is worth to note that strong sc-entailment from a conjunction of wffs and strong sc-entailment from a set of all its conjuncts only overlap, but not coincide. Clearly, the following is true:

Corollary 33. *Let $X \neq \emptyset$. If $X \mathrel{\triangleright\!\sim} B$, then $\bigwedge X \mathrel{\triangleright\!\sim} B$.*

For example, $\{p, q\} \mathrel{\triangleright\!\sim} p \wedge q$ is the case and thus $p \wedge q \mathrel{\triangleright\!\sim} p \wedge q$ holds as well. Yet, the converse of Corollary 33 is not true. For instance, $p \wedge q \mathrel{\triangleright\!\sim} p$ holds, while $\{p, q\} \mathrel{\triangleright\!\sim} p$ does not hold.[12] At first sight this looks untenable. However, the phenomenon can be explained as follows. Information carried by $\bigwedge X \mathrel{\triangleright\!\sim} B$ and $X \mathrel{\triangleright\!\sim} B$ differ when X is not a singleton set. In both cases transmission of truth as well as consistency of the set X are ensured. The claim of $\bigwedge X \mathrel{\triangleright\!\sim} B$ is: although B need not be true, the (hypothetical) truth of all the wffs in X is sufficient for B be true. Note that $\bigwedge X \mathrel{\triangleright\!\sim} B$ does not exclude that the transmission of truth effect takes place w.r.t. some proper subset or some proper superset of X. (As for $p \wedge q \mathrel{\triangleright\!\sim} p$, there is a proper subset of $\{p, q\}$, namely $\{p\}$, which ensures the transmission.) The claim of $X \mathrel{\triangleright\!\sim} B$ is stronger: this is just the (hypothetical) truth of all the wffs in X that warrants the (hypothetical) truth of B. "Just" means here: "one needs neither more nor less than the truth of *all* the wffs in X for B be true."

Observe that one can pass from $\bigwedge X \mathrel{\triangleright\!\sim} B$ to $X \mathrel{\triangleright\!\sim} B$ on the condition:

($\heartsuit$) *for each $A \in X : \bigwedge(X_{\ominus A}) \not\models B$*

which, however, does not hold universally.

---

[12] By the way, these examples provide a nice illustration of the lack of transitivity of strong sc-entailment.

Remark 5. Although strong sc-entailment is "antimonotone" (cf. Corollary 15), the following fact is worth some attention:

Corollary 34. *Let $X \neq \emptyset$. If $X \not\Vdash B$ and Y is a consistent proper superset of X, then $\bigwedge Y \not\Vdash B$.*

Proof. By Corollary 19, if $X \neq \emptyset$ and $X \not\Vdash B$, then B is not valid. So $\emptyset \not\models B$. On the other hand, $\emptyset$ is the only proper subset of $\{\bigwedge Y\}$. Clearly, if $X \models B$, then $Y \models B$ and hence $\{\bigwedge Y\} \models B$. If Y is consistent, so is $\{\bigwedge Y\}$. Therefore $\bigwedge Y \not\Vdash B$. □

Thus a wff strongly sc-entailed by a non-empty set of wffs X is also strongly sc-entailed by (the singleton set comprising) a conjunction of all the wffs of a consistent extension Y of X. Note, however, that, according to what has been said above, $\bigwedge Y \not\Vdash B$ carries less information than $X \not\Vdash B$. Moreover, $X \not\Vdash B$ suppresses $Y \not\Vdash B$.

5 Some Comparisons

5.1 Strong vs. Classical

The basic properties of strong entailments differ from those of their classical counterparts. However, one can show that whatever is reachable by classical entailments from consistent sets of premises, is also attainable by strong entailments from some finite subsets of these sets. To put it briefly: no classical consequence of a consistent set is lost.

Notice that it holds that (we present a proof of this well-known fact only to keep this paper self-contained):

Lemma 1. *Each inconsistent set of wffs has a subset being a MI-set.*

Proof. Let X be an inconsistent set of wffs. By compactness of CPL, X has an inconsistent finite subset, say, X'. Clearly, $X' \neq \emptyset$. Consider the family of all inconsistent subsets of X'. Let us designate it by Ψ. Since X' is inconsistent, $\Psi \neq \emptyset$. As X' is non-empty and finite, there is a natural number, say, k, where $k \geqslant 1$, such that no set in Ψ has less than k elements. Let Y be an element of Ψ which comprises exactly k wffs. Obviously, no proper subset of Y belongs to Ψ. Therefore each proper subset of Y is consistent. It follows that Y is a MI-set included in X. □

Let us now prove:

Theorem 10 (Simulation of $\models$). *If $X \models Y$ and X is consistent, then there exist a finite subset X_1 of X and a finite non-empty subset Y_1 of Y such that $X_1 \Vdash Y_1$.*

Proof. If $X \models Y$, then the set $X, \neg Y$ is inconsistent and thus, by Lemma 1, has subset(s) being MI-sets. Let Z be a MI-set such that $Z \subseteq X, \neg Y$. Since, by assumption, X is consistent, $Z \not\subseteq X$. We put:

$$X_1 =_{df} X \cap Z$$
$$W =_{df} Z \setminus X_1$$

Clearly, $W \subseteq \neg Y$. Moreover, $W \neq \emptyset$, and $Z = X_1, W$ as well as $X_1 \cap W = \emptyset$. Consider the set Y_1 defined by:

$$Y_1 =_{df} \{C : \neg C \in W\}.$$

We have $W = \neg Y_1$ and hence $Z = X_1, \neg Y_1$. It follows that $Y_1 \subseteq Y$ and $X_1 \cap \neg Y_1 = \emptyset$. Since Z is a MI-set and $Z = X_1, \neg Y_1$ as well as $X_1 \cap \neg Y_1 = \emptyset$, by Theorem 5 we conclude that $X_1 \Vdash Y_1$ holds. As each MI-set is finite, X_1 and Y_1 are finite subsets of X and Y, respectively. Finally, $Y_1 \neq \emptyset$ since $W \neq \emptyset$. □

As a consequence of Definition 6 and Theorem 10 we get:

Theorem 11 (Simulation of $\models$). *If $X \models B$ and X is consistent, then there exists a finite subset Z of X such that $Z \Vdash B$.*

Proof. Recall that $X \models B$ iff $X \models \{B\}$, and $Z \Vdash B$ iff $Z \Vdash \{B\}$. Since we have already proven Theorem 10, it suffices to observe that the only non-empty subset of the singleton set $\{B\}$ is $\{B\}$ itself. □

The intuitive content of Theorem 11 is this: CPL sc-entailment from a given, finite or infinite, consistent set of wffs boils down to strong sc-entailment from a finite subset of the set. Theorem 10 presents an analogous result for mc-entailment.

Remark 6. Let X and Y be different, yet logically equivalent consistent sets of wffs. The set of wffs classically sc-entailed by X coincides with the set of wffs classically entailed by Y. However, this need not be the case for strong sc-entailment. Yet, Theorem 11 yields that the set of wffs attainable by strong sc-entailment from some finite subset of X equals the set of wffs which are obtainable by strong sc-entailment from some finite subset of Y, and equals the set comprising all the wffs classically sc-entailed by X or by Y. Needless to say, the respective subsets of X and of Y may differ.

5.2 Strong vs. Relevant

As it is well-known, when the sum of two consistent sets of CPL-wffs, X and Y, is inconsistent, then $\text{Var}(X) \cap \text{Var}(Y) \neq \emptyset$ (cf. e.g. [6], p. 375). It follows that classical sc-entailment from consistent sets of premises to conclusions which are not valid wffs exhibits the variable sharing property. It is worth to note that the same holds true for strong sc-entailment. Theorem 11 together with corollaries 24 and 20 almost immediately yield:

Corollary 35. *Let X be a non-empty, consistent set of wffs. If $X \models B$ and B is not a valid wff, then there exists a finite, non-empty subset Z of X such that $Z \mathrel{\vdash\!\!\!\mid} B$ and $\text{Var}(Z) \cap \text{Var}(B) \neq \emptyset$.*

Variable sharing is often regarded as an indicator (or even a precondition) of relevance in the context of semantic consequence. As such, it is usually invoked in relevant logics. So the question arises: what is the relation between strong sc-entailment and accounts of entailment proposed in relevant logics? Since there exist many systems of relevance logic, an exhaustive answer would have required a separate paper. For the reasons of space, let me restrict to a few remarks only.

As for CPL, valid wffs falling under the schema:

$$A \to B \qquad (31)$$

license sc-entailment of B from A. For the lack of a better idea, let us call them *classical implicational laws* or briefly CIL's.[13]

Recall that although all classically valid wffs are strongly sc-entailed by the empty set, the transition from $\emptyset \mathrel{\vdash\!\!\!\mid} A \to B$ to $A \mathrel{\vdash\!\!\!\mid} B$ is not always legitimate (cf. Corollary 28). Corollary 30 yields, in turn, that a CIL does not license strong sc-entailment just in case its antecedent or consequent is not contingent.

The first observation is: there exist CIL's which are both rejected in some relevant logics[14] and do not license strong sc-entailment. Examples are shown in Table 1.

Second, there exist CIL's which are rejected in some relevant logic(s), but license strong sc-entailment. Examples are given in Table 2.

Third, it happens that a CIL which is accepted in a relevant logic does not license strong sc-entailment. The "mingle" formulas, i.e. wffs of the form $A \to (A \to A)$, provide simple examples here.

---

[13] One should not confuse CIL's with laws of the implicational fragment of CPL. Both A and B may involve any connective, implication included. What is important is that implication is the main connective of (the wff which expresses) a CIL.

[14] That is, at least one of well-known relevant logics rejects the corresponding law; there is no space for details. For relevant logics see, e. g., [13] and [18].

$p \wedge \neg p \to q$	$p \wedge \neg p \not\Kappa q$
$\neg(p \to p) \to q$	$\neg(p \to p) \not\Kappa q$
$p \to (q \to q)$	$p \not\Kappa q \to q$
$p \to (p \to p)$	$p \not\Kappa p \to p$
$p \to p \vee \neg p$	$p \not\Kappa p \vee \neg p$
$p \to q \vee \neg q$	$p \not\Kappa q \vee \neg q$
$(p \to p) \to (q \to q)$	$p \to p \not\Kappa q \to q$
$(p \to q) \to (p \to p)$	$p \to q \not\Kappa p \to p$

Table 1: *Examples of* CIL*'s rejected in some relevant logics (left column) which do not license strong sc-entailment (as depicted in the right column).*

$p \to (q \to p)$	$p \Kappa (q \to p)$
$p \to (\neg p \to q)$	$p \Kappa (\neg p \to q)$
$p \to ((p \to q) \to q)$	$p \Kappa (p \to q) \to q$
$((p \to q) \to p) \to p$	$(p \to q) \to p \Kappa p$
$(p \to (q \to r)) \to (q \to (p \to r))$	$p \to (q \to r) \Kappa q \to (p \to r)$
$p \wedge q \to (p \to q) \wedge (q \to p)$	$p \wedge q \Kappa (p \to q) \wedge (q \to p)$
$p \wedge (\neg p \vee q) \to q$	$p \wedge (\neg p \vee q) \Kappa q$

Table 2: *Examples of* CIL*'s rejected in some relevant logics (left column) which, however, license strong sc-entailment (as depicted in the right column).*

5.3 Strong vs. Connexive

Connexive logics are usually characterized as systems validating the following theses:[15]

$$\neg(\neg A \to A) \tag{32}$$

$$\neg(A \to \neg A) \tag{33}$$

$$(A \to B) \to \neg(A \to \neg B) \tag{34}$$

$$(A \to \neg B) \to \neg(A \to B) \tag{35}$$

As strong sc-entailment from the empty set is restricted to classically valid wffs only (cf. Corollary 20) and some wffs of the forms (32 – 35) are not classically valid, it is not the case that all the wffs falling under the schemata (32) – (35) are strongly

---

[15]For connexive logics see, e.g. [14]. Theses (32) and (33) are attributed to Aristotle, while theses (34) and (35) are ascribed to Boethius.

sc-entailed by the empty set.[16] It is worth to note, however, that the following are true:

Corollary 36. *For any wff A:*

1. $\neg A \not\hspace{-2pt}\mid\hspace{-4pt}\prec A$,

2. $A \not\hspace{-2pt}\mid\hspace{-4pt}\prec \neg A$.

Proof. Suppose that $\neg A \mid\hspace{-4pt}\prec A$ for some wff A. Thus $\neg A \models A$. On the other hand, by Theorem 1 it follows that both $\neg A$ and A are contingent wffs. Hence $\neg A \not\models A$.
We reason analogously in the case of (*2*). □

Thus a negation of a wff never strongly sc-entails the wff itself, and a wff never strongly sc-entails its negation. Corollary 36 seems to express an idea akin to that which lies behind having (32) and (33) as theses.

Corollary 37. *For any wffs A, B:*

1. *if* $A \mid\hspace{-4pt}\prec B$, *then* $A \not\hspace{-2pt}\mid\hspace{-4pt}\prec \neg B$,

2. *if* $A \mid\hspace{-4pt}\prec \neg B$, *then* $A \not\hspace{-2pt}\mid\hspace{-4pt}\prec B$,

Proof. Assume that $A \mid\hspace{-4pt}\prec B$. It follows that A is a consistent wff and $A \models B$. Therefore there exists a valuation v such that $v(A) = \mathbf{1}$, $v(B) = \mathbf{1}$ and hence $v(\neg B) = \mathbf{0}$. Thus $A \not\models \neg B$. It follows that $A \not\hspace{-2pt}\mid\hspace{-4pt}\prec \neg B$.
We reason similarly in the case of (*2*). □

A due comment on Corollary 37 is analogous to that on Corollary 36.

6 Towards a Proof-theoretic Account of Strong Entailments

As Theorem 5 shows, a problem of the form:

(P) *Does X strongly mc-entail Y?*

splits into two sub-problems:

(P$_1$) *Is it the case that $X \cap \neg Y = \emptyset$?*

---

[16]However, some of them are classically valid and thus *are* strongly sc-entailed by the empty set. For instance, we have $\emptyset \mid\hspace{-4pt}\prec \neg(\neg(p \wedge \neg p) \rightarrow p \wedge \neg p)$, $\emptyset \mid\hspace{-4pt}\prec \neg((p \rightarrow p) \rightarrow \neg(p \rightarrow p))$, or $\emptyset \mid\hspace{-4pt}\prec (p \vee \neg p \rightarrow p \wedge \neg p) \rightarrow \neg(p \vee \neg p \rightarrow \neg(p \wedge \neg p))$.

(**P$_2$**) *Is $X, \neg Y$ a MI-set?*

Similarly, due to Theorem 8, a problem of the form:

(**P'**) *Does X strongly sc-entail B?*

splits into:

(**P'$_1$**) *Is it the case that '$\neg B$' $\notin X$?*

(**P'$_2$**) *Is $X, \neg B$ a MI-set?*

P$_1$ and **P'$_1$** are purely syntactic issues. But either **P$_2$** or **P'$_2$** is a problem that pertains to a semantic property. In order to solve it syntactically one needs a proof-theoretic account of MI-sets.

6.1 The Calculus MI^{CPL}

In this section we present a calculus, labelled MI^{CPL}, in which provable sequents of a strictly defined form correspond to MI-sets.

Rules of the calculus operate on sequences of sequents of a specific kind. Since a sequence of sequents is customarily called a hypersequent, MI^{CPL} may be called a calculus of hypersequents. But speaking about hypersequent calculi usually brings into mind Avron's seminal works.[17] However, the format of MI^{CPL} differs considerably from that of Avron-style hypersequent calculi. In particular, derivations and proofs in MI^{CPL} are not trees having hypersequents in their nodes, but sequences of hypersequents. Rules of MI^{CPL} transform hypersequents into hypersequents, and a rule is always applied to the last term of a derivation constructed so far. Last but not least, MI^{CPL} has no axioms, but comprises rules only.

Given these substantial differences, and taking into account that the concept of hypersequent is loaded with references to Avron-style calculi, let me use a new term for a sequence of sequents. The term chosen is "seqsequent", after Latin *sequentia*, which means (among others) "sequence."[18] No doubt, saying that we aim at a calculus of seqsequents is less misleading than speaking about a calculus of hypersequents. A warning is needed, however. As we will see, the order in which

---

[17]Starting from the influential paper [1]. Avron-style approach is not the only one, however. A reader interested in different types of hypersequent calculi (including those in which hypersequents are construed as sets or multisets of sequents rather than their sequences) is advised to consult [8], Chapter 4.7.

[18]"Seq" is not a prefix in English, but since many English words begin with prefixes rooted in Latin, I hope that this proposal is acceptable, at least for the purposes of this paper. A reader familiar with programming is kindly requested to suspend any associations he/she may have.

sequents occur in a "seqsequent" does not determine the order of application of rules of the calculus. We are speaking about a calculus of seqsequents only to stress that rules of the calculus operate on "seqsequents", that is, a rule transforms a sequence of sequents into a sequence of sequents.

6.1.1 Numerically Annotated Wffs, Sequents, and Seqsequents

We will be operating with sequents based on sequences of numerically annotated wffs. By a *numerically annotated wff* (na-wff for short) we mean an expression of the form $A^{[i]}$, where A is a wff and i is a numeral from the set $\{1, 2, 3, \ldots\}$. Let us stress that numerals are here proof-theoretic devices only. It is not assumed that they refer to possible worlds or perform the function of labels.

By a *sequent* we will mean an expression of the form:

$$C_1^{[i_1]}, \ldots, C_m^{[i_m]} \vdash \tag{36}$$

where $C_1^{[i_1]}, \ldots, C_m^{[i_m]}$ is a finite sequence of na-wffs; when $m = 0$, we write the corresponding sequent as $\emptyset \vdash$. Although we consider sequents with empty succedents, it is no accident that we put the turnstile $\vdash$ into a sequent. This will allow us to differentiate between operations on sequents and operations on sequences of annotated wffs (see below).

An *atomic sequent* is of the form:

$$l_1^{[i_1]}, \ldots, l_m^{[i_m]} \vdash \tag{37}$$

where $l_1, \ldots, l_m$ are literals, that is, propositional variables or their negations. An atomic sequent (37) is *closed* if it involves na-wffs based on complementary literals, i.e. there exist $l_j^{[i_j]}, l_k^{[i_k]}$ ($1 \leqslant j, k \leqslant m$) such that $l_j = \neg l_k$. An atomic sequent which is not closed is called *open*.

We use the Greek lower-case letters σ, θ, χ, possibly with subscripts, as metalanguage variables for finite sequences of na-wffs, the empty sequence included.

Let $\sigma \vdash$ be a sequent. We define:

$$\mathsf{wff}(\sigma \vdash) = \{A \in \texttt{Form} : A^{[i]} \text{ is a term of } \sigma\}.$$

By $f_{\setminus [i_j]}(\sigma)$ we mean the subsequence of σ resulting from it by removing all its terms (i.e. na-wffs) which are annotated with the numeral i_j. Needless to say, $f_{\setminus [i_j]}(\sigma) \vdash$ is a sequent.[19]

---

[19] If i_j is the only numeral which occurs in na-wffs of σ, then $f_{\setminus [i_j]}(\sigma) \vdash$ equals $\emptyset \vdash$, which is, by definition, a sequent. Of course, $\mathsf{wff}(\emptyset \vdash) = \emptyset$.

A *seqsequent* is a finite sequence of sequents. We use the Greek upper-case letters Φ, Ψ, Θ, with subscripts when necessary, as metalinguistic variables for seqsequents. By a *constituent* of a seqsequent we mean any sequent which is a term of the seqsequent.

Finally, we distinguish *ordered sequents*.

An ordered sequent is a sequent which falls under the schema:

$$C_1^{[1]}, \ldots, C_m^{[m]} \vdash \tag{38}$$

where $m \geqslant 1$, and $C_1, \ldots, C_m$ are pairwise syntactically distinct wffs when $m > 1$. Thus, besides sequents of the form $A^{[1]} \vdash$, ordered sequents are sequents whose consecutive terms (with the exception of the turnstile), are pairwise syntactically distinct wffs annotated with consecutive numerals (occurring in curly brackets), starting from the numeral 1.[20] At the metalanguage level, ordered sequents of the form (38) will be concisely written as:

$$C_{\vec{m}}^{[\vec{m}]} \vdash \tag{39}$$

As we will see, in order to show that $\{C_1, \ldots, C_m\}$ is a MI-set it suffices to prove the corresponding ordered sequent $C_{\vec{m}}^{[\vec{m}]} \vdash$. Moreover, having a disproof of the ordered sequent is tantamount to showing that the corresponding set of wffs, albeit inconsistent, is not a MI-set.

6.1.2 Rules and Proofs

In order to present the rules of $\mathsf{MI}^{\mathsf{CPL}}$ in a concise manner let us introduce some notational conventions first.

Following [22], we distinguish between α-wffs and β-wffs, and we assign two further wffs to any of them, in the way presented in Table 3.

We use the sign ' as the concatenation-sign for sequences of na-wffs. For brevity, we assume that a metalanguage expression of the form $\sigma \, ' A^{[i]}$ denotes the concatenation of sequence σ and the one-term sequence $\langle A^{[i]} \rangle$, while a metalanguage expression of the form $\sigma \, ' A^{[i]} \, ' \theta$ refers to the concatenation of sequence $\sigma \, ' A^{[i]}$ and sequence θ.

The semicolon will perform the role of the concatenation-sign for seqsequents. We usually omit angle brackets when referring to a seqsequent which has only one constituent. Thus $\Psi; \sigma \vdash$ stands for the concatenation of Ψ and $\langle \sigma \vdash \rangle$. The expression $\Psi \, ; \sigma \vdash; \Phi$ refers to the concatenation of $\Psi; \sigma \vdash$ and Φ.

---

[20]Each ordered sequent is a sequent, but not the other way round. For instance, the expressions $p^{[4]}, q^{[2]} \vdash$ and $p^{[1]}, p^{[3]} \vdash$ are sequents in our sense, but none of them is an ordered sequent. Similarly, $p^{[1]}, p^{[1]} \vdash$ and $p^{[1]}, p^{[2]} \vdash$ are sequents, though neither of them is an ordered sequent.

α	α_1	α_2	β	β_1	β_2
$A \wedge B$	A	B	$\neg(A \wedge B)$	$\neg A$	$\neg B$
$\neg(A \vee B)$	$\neg A$	$\neg B$	$A \vee B$	A	B
$\neg(A \to B)$	A	$\neg B$	$A \to B$	$\neg A$	B

Table 3: α-wffs and β-wffs

The calculus MI^{CPL} has only rules which operate on seqsequents. No axioms are provided. Here are the primary rules of MI^{CPL}:

$$R_\alpha^{[i]} : \quad \frac{\Phi;\ \sigma\ '\ \alpha^{[i]}\ '\ \theta \vdash; \Psi}{\Phi;\ \sigma\ '\ \alpha_1^{[i]}\ '\ \alpha_2^{[i]}\ '\ \theta \vdash; \Psi}$$

$$R_\beta^{[i]} : \quad \frac{\Phi;\ \sigma\ '\ \beta^{[i]}\ '\ \theta \vdash; \Psi}{\Phi;\ \sigma\ '\ \beta_1^{[i]}\ '\ \theta \vdash;\ \sigma\ '\ \beta_2^{[i]}\ '\ \theta \vdash; \Psi}$$

$$R_{\neg\neg}^{[i]} : \quad \frac{\Phi;\ \sigma\ '\ \neg\neg A^{[i]}\ '\ \theta \vdash; \Psi}{\Phi;\ \sigma\ '\ A^{[i]}\ '\ \theta \vdash; \Psi}$$

Any of Φ, Ψ, σ, θ can be empty.

Observe that rules of MI^{CPL} "act locally": if a rule is applied to a seqsequent, only one constituent and only one occurrence of a na-wff in the constituent are acted upon, while the other occurrences and other constituents remain unaffected. Moreover, any new na-wff that comes into play due to an application of a rule is annotated with the same numeral as the na-wff acted upon.

We are now ready for an introduction of the concept of proof.

Definition 7 (Proof). *A finite sequence of seqsequents $\Theta_1, \ldots, \Theta_n$ is a MI^{CPL}-proof of an ordered sequent $C_{\vec{m}}^{[\vec{m}]} \vdash$ iff*

1. $\Theta_1 = \langle C_{\vec{m}}^{[\vec{m}]} \vdash \rangle$,

2. Θ_{j+1} *results from Θ_j by a rule of* MI^{CPL}, *where $1 \leqslant j < n$,*

3. *each constituent of Θ_n is a closed atomic sequent,*

4. *for each $k \in \{1, \ldots, m\}$ there exists a constituent $\sigma \vdash$ of Θ_n such that the sequent $f_{\setminus [k]}(\sigma) \vdash$ is an open atomic sequent or is of the form $\emptyset \vdash$.*

An ordered sequent is provable in MI^{CPL} *iff the sequent has a* MI^{CPL}*-proof.*

Remark 7. The concept of proof introduced above is non-standard in many respects. First, a proof is a sequence of seqsequents. Second, notice that it is an ordered sequent (i.e. a sequent in which wffs occurring left of the turnstile are annotated with consecutive numerals, starting from 1) that performs the role of an "input" of a proof: the first line of a proof is a one-term seqsequent involving an ordered sequent. We do not introduce the concept of proof of a sequent in general, but only of an ordered sequent. As we will see, this is sufficient for our purposes. Third, proofs in MI^{CPL} are strictly linear: Θ_{j+1} results by a rule from Θ_j only. Clauses (3) and (4) of the definition jointly ensure that $\mathsf{wff}(C_{\vec{m}}^{[\vec{m}]} \vdash)$ is a MI-set.

Provability in MI^{CPL} and the property of being a MI-set are linked in a way characterized by:

Theorem 12 (Soundness w.r.t. MI-sets). *Let X be a finite non-empty set of wffs, and let $\sigma \vdash$ be an ordered sequent such that $\mathsf{wff}(\sigma \vdash) = X$. If the sequent $\sigma \vdash$ is provable in* MI^{CPL}*, then X is a MI-set.*

A proof of Theorem 12 is presented in the Appendix.

Due to Theorem 12, in order to show that X is a MI-set it suffices to prove an ordered sequent $C_{\vec{m}}^{[\vec{m}]} \vdash$ for which the equation $X = \mathsf{wff}(C_{\vec{m}}^{[\vec{m}]})$ holds.

Example 1. $\{p, \neg p\}$ *is a* MI*-set.*

The one-term sequence $\langle p^{[1]}, \neg p^{[2]} \vdash \rangle$ is a proof of the sequent $p^{[1]}, \neg p^{[2]} \vdash$, since $f_{\backslash [1]}(p^{[1]}, \neg p^{[2]}) \vdash\; =\; \neg p^{[2]} \vdash$ and $f_{\backslash [2]}(p^{[1]}, \neg p^{[2]}) \vdash\; =\; p^{[1]} \vdash$.

For brevity, in what follows we will be omitting angle brackets in the first line of a proof, and in the case of one-term seqsequents.

Example 2. $\{p \wedge \neg p\}$ *is a* MI*-set.*

Here is a proof of the sequent $(p \wedge \neg p)^{[1]} \vdash$ (inscriptions of the form $\mathsf{R}_x^{[i]}$ do not belong to proofs, but indicate what rule has been applied to the seqsequent which occurs on the left):

$(p \wedge \neg p)^{[1]} \vdash \qquad \mathsf{R}_\alpha^{[1]}$
$p^{[1]}, \neg p^{[1]} \vdash$

Notice that $f_{\backslash [1]}(p^{[1]}, \neg p^{[1]} \vdash)$ equals $\emptyset \vdash$.

Example 3. $\{p, \neg(\neg p \to q)\}$ *is a* MI*-set.*

The following is a proof of a corresponding ordered sequent:

$p^{[1]}, \neg(\neg p \to q)^{[2]} \vdash \quad\quad\quad \mathsf{R}_\alpha^{[2]}$
$p^{[1]}, \neg p^{[2]}, \neg q^{[2]} \vdash$

For the sake of transparency, from now on we highlight the na-wff on which the rule indicated to the right acts upon. We tick exemplary occurrences of numerals due to which clause (4) of the definition of proof is satisfied.

Example 4. $\{\neg p, \neg q, p \vee q\}$ *is a MI-set.*

Here is a proof of a corresponding ordered sequent:

$\neg p^{[1]}, \neg q^{[2]}, \, (p \vee q)^{[3]} \vdash \quad\quad\quad \mathsf{R}_\beta^{[3]}$
$\neg p^{[1\checkmark]}, \neg q^{[2]}, p^{[3\checkmark]} \vdash; \neg p^{[1]}, \neg q^{[2\checkmark]}, q^{[3]} \vdash$

Example 5. $\{p \vee (q \vee r), \neg(p \vee q), \neg(p \vee r)\}$ *is a MI-set.*

$p \vee (q \vee r))^{[1]}, \neg(p \vee q)^{[2]}, \neg(p \vee r)^{[3]} \vdash \quad\quad\quad \mathsf{R}_\alpha^{[3]}$
$(p \vee (q \vee r))^{[1]}, \; \neg(p \vee q)^{[2]}, \neg p^{[3]}, \neg r^{[3]} \vdash \quad\quad\quad \mathsf{R}_\alpha^{[2]}$
$(p \vee (q \vee r))^{[1]}, \neg p^{[2]}, \neg q^{[2]}, \neg p^{[3]}, \neg r^{[3]} \vdash \quad\quad\quad \mathsf{R}_\beta^{[1]}$
$p^{[1]}, \neg p^{[2]}, \neg q^{[2]}, \neg p^{[3]}, \neg r^{[3]} \vdash; (q \vee r)^{[1]}, \neg p^{[2]}, \neg q^{[2]}, \neg p^{[3]}, \neg r^{[3]} \vdash \quad \mathsf{R}_\beta^{[1]}$
$p^{[1\checkmark]}, \neg p^{[2]}, \neg q^{[2]}, \neg p^{[3]}, \neg r^{[3]} \vdash; q^{[1]}, \neg p^{[2]}, \neg q^{[2\checkmark]}, \neg p^{[3]}, \neg r^{[3]} \vdash;$
$\quad\quad\quad\quad\quad r^{[1]}, \neg p^{[2]}, \neg q^{[2]}, \neg p^{[3]}, \neg r^{[3\checkmark]} \vdash$

Example 6. $\{p \to (q \to r), p \to q, \neg(p \to r)\}$ *is a MI-set.*

$(p \to (q \to r))^{[1]}, (p \to q)^{[2]}, \neg(p \to r)^{[3]} \vdash \quad\quad\quad \mathsf{R}_\alpha^{[3]}$
$(p \to (q \to r))^{[1]}, (p \to q)^{[2]}, p^{[3]}, \neg r^{[3]} \vdash \quad\quad\quad \mathsf{R}_\beta^{[2]}$
$(p \to (q \to r))^{[1]}, \neg p^{[2]}, p^{[3]}, \neg r^{[3]} \vdash; (p \to (q \to r))^{[1]}, q^{[2]}, p^{[3]}, \neg r^{[3]} \vdash \quad \mathsf{R}_\beta^{[1]}$
$\neg p^{[1]}, \neg p^{[2]}, p^{[3]}, \neg r^{[3]} \vdash; (q \to r)^{[1]}, \neg p^{[2]}, p^{[3]}, \neg r^{[3]} \vdash;$
$\quad\quad\quad\quad\quad (p \to (q \to r))^{[1]}, q^{[2]}, p^{[3]}, \neg r^{[3]} \vdash \quad \mathsf{R}_\beta^{[1]}$
$\neg p^{[1]}, \neg p^{[2]}, p^{[3]}, \neg r^{[3]} \vdash; \neg q^{[1]}, \neg p^{[2]}, p^{[3]}, \neg r^{[3]} \vdash; r^{[1]}, \neg p^{[2]}, p^{[3]}, \neg r^{[3]} \vdash;$
$\quad\quad\quad\quad\quad (p \to (q \to r))^{[1]}, q^{[2]}, p^{[3]}, \neg r^{[3]} \vdash \quad \mathsf{R}_\beta^{[1]}$
$\neg p^{[1]}, \neg p^{[2]}, p^{[3]}, \neg r^{[3]} \vdash; \neg q^{[1]}, \neg p^{[2]}, p^{[3]}, \neg r^{[3]} \vdash; r^{[1]}, \neg p^{[2]}, p^{[3]}, \neg r^{[3]} \vdash;$
$\quad\quad\quad\quad\quad \neg p^{[1]}, q^{[2]}, p^{[3]}, \neg r^{[3]} \vdash; (q \to r)^{[1]}, q^{[2]}, p^{[3]}, \neg r^{[3]} \vdash \quad \mathsf{R}_\beta^{[1]}$
$\neg p^{[1]}, \neg p^{[2]}, p^{[3\checkmark]}, \neg r^{[3]} \vdash; \neg q^{[1]}, \neg p^{[2\checkmark]}, p^{[3]}, \neg r^{[3]} \vdash; r^{[1]}, \neg p^{[2]}, p^{[3]}, \neg r^{[3]} \vdash;$

$\neg p^{[1^\vee]}, q^{[2]}, p^{[3]}, \neg r^{[3]} \vdash; \neg q^{[1]}, q^{[2]}, p^{[3]}, \neg r^{[3]} \vdash; r^{[1]}, q^{[2]}, p^{[3]}, \neg r^{[3]} \vdash$

The system MI^{CPL} is complete w.r.t. MI-sets.

Theorem 13 (Completeness w.r.t. MI-sets). *If X is a MI-set, then any ordered sequent $\sigma \vdash$ such that $\mathsf{wff}(\sigma \vdash) = X$ is provable in MI^{CPL}.*

A proof of Theorem 13 is presented in the Appendix.

6.1.3 Disproofs

The system MI^{CPL} is useful not only in showing that something is a MI-set, but also in demonstrating that a set of wffs is inconsistent yet not minimally so. The latter can be achieved by providing a *disproof* of an ordered sequent which corresponds to the set of wffs under consideration.

Definition 8 (Disproof). *A finite sequence of seqsequents $\Theta_1, \ldots, \Theta_n$ is a MI^{CPL}-disproof of an ordered sequent $C_{\vec{m}}^{[\vec{m}]} \vdash$ iff*

1. *$\Theta_1 = \langle C_{\vec{m}}^{[\vec{m}]} \vdash \rangle$,*

2. *Θ_{j+1} results from Θ_j by a rule of MI^{CPL}, where $1 \leqslant j < n$,*

3. *each constituent of Θ_n is a closed atomic sequent,*

4. *there exists $k \in \{1, \ldots, m\}$ such that for each constituent $\sigma \vdash$ of Θ_n, the sequent $f_{\backslash [k]}(\sigma) \vdash$ is closed.*

An ordered sequent is disprovable in MI^{CPL} iff the sequent has a MI^{CPL}-disproof.

Observe that proofs and disproofs differ only with respect to their closing conditions. To be more precise, each sequence of seqsequents $\Theta_1, \ldots, \Theta_n$ satisfying the clauses 1, 2, and 3 of the definition of proof (i.e. Definition 7) and violating clause 4 of the definition is not a proof, but a disproof.

The following holds (for a proof, see the Appendix):

Theorem 14. *If there exists a MI^{CPL}-disproof of an ordered sequent $C_{\vec{m}}^{[\vec{m}]} \vdash$, then the set $\mathsf{wff}(C_{\vec{m}}^{[\vec{m}]} \vdash)$ is inconsistent, but is not a MI-set.*

Here is an example of a disproof of $(p \to q)^{[1]}, p^{[2]}, \neg(p \vee q)^{[3]} \vdash$ (clause 4 is satisfied w.r.t. the numeral 1):

Example 7. $(p \to q)^{[1]}, p^{[2]}, \neg(p \vee q)^{[3]} \vdash$ $\qquad$ $\mathsf{R}^{[2]}_\beta$

$\neg p^{[1]}, p^{[2]}, \neg(p \vee q)^{[3]} \vdash; q^{[1]}, p^{[2]}, \neg(p \vee q)^{[3]} \vdash$ $\qquad$ $\mathsf{R}^{[2]}_\alpha$

$\neg p^{[1]}, p^{[2]}, \neg p^{[3]}, \neg q^{[3]} \vdash; q^{[1]}, p^{[2]}, \neg(p \vee q)^{[3]} \vdash$ $\qquad$ $\mathsf{R}^{[2]}_\alpha$

$\neg p^{[1]}, p^{[2]}, \neg p^{[3]}, \neg q^{[3]} \vdash; q^{[1]}, p^{[2]}, \neg p^{[3]}, \neg q^{[3]} \vdash$

Thus $\{p \to q, p, \neg(p \vee q)\}$, though inconsistent, is not a MI-set.

Finally, the following holds as well:

Theorem 15. *If X is a finite inconsistent set of wffs which is not a MI-set, then any ordered sequent $\sigma \vdash$ such that $\mathsf{wff}(\sigma \vdash) = X$ is disprovable in $\mathsf{MI}^{\mathsf{CPL}}$.*

For a proof of Theorem 15 see the Appendix.

Remark 8. The primary rules of $\mathsf{MI}^{\mathsf{CPL}}$ transform wffs inside sequents analogously as Smullyan's tableaux rules do. It is possible to build a calculus of MI-sets in the "standard" tableau format, with rules defined as operating directly on (annotated) wffs, while occurrences of these wffs are nodes of respective trees. This would require adding an annotation mechanism and specifying new closing conditions. The format of $\mathsf{MI}^{\mathsf{CPL}}$ is akin to that of the so-called erotetic calculi (cf., e.g.,[27], [11],[10]). The main difference lies in the fact that rules of $\mathsf{MI}^{\mathsf{CPL}}$ operate on sequences of sequents, while rules of erotetic calculi act upon questions based on sequences of sequents. Moreover, annotations are exploited here in a new manner, and closing conditions of a proof are more demanding. The advantage of the current format over the "standard" tableaux approach lies in its relative simplicity at the metatheoretical level. Moreover, it is known that proofs written in the erotetic calculi format can be transformed into proofs in tableaux calculi (cf.[10]), sequent calculi (cf. [11]) or even Hilbert-style calculi (cf. [7]). These effects do not disappear when we move from questions based on sequences of sequents to the "inner" sequences of sequents.

6.2 Soundness and Completeness of $\mathsf{MI}^{\mathsf{CPL}}$ w.r.t. Strong Entailments

As we have shown, the calculus $\mathsf{MI}^{\mathsf{CPL}}$ is sound and complete w.r.t. MI-sets. Due to Theorem 5, the fact that $X, \neg Y$ is a MI-set guarantees that $X \parallel\!\prec Y$ holds provided that $X \cap \neg Y = \emptyset$ is the case. So when we restrict ourselves to ordered sequents built in such a way that the fulfilment of the latter condition is secured, proofs of these ordered sequents can be viewed as demonstrations that strong mc-entailment hold in the cases considered.

Theorem 16 (Soundness w.r.t. strong mc-entailment). *Let $X = \{A_1, \ldots, A_n\}$ and $Y = \{B_1, \ldots, B_k\}$, where $n + k > 0$ and $A_i \neq \neg B_j$ for $i = 1, \ldots, n$ and $j = 1, \ldots, k$. If the ordered sequent:*

$$A_1^{[1]}, \ldots, A_n^{[n]}, \neg B_1^{[n+1]}, \ldots, \neg B_k^{[n+k]} \vdash$$

is provable in $\mathsf{MI^{CPL}}$*, then $X \Vdash\!\!\!\prec Y$.*

Proof. By Theorem 5 and Theorem 12. □

Theorem 17 (Completeness w.r.t. strong mc-entailment). *Let $X = \{A_1, \ldots, A_n\}$ and $Y = \{B_1, \ldots, B_k\}$, where $n + k > 0$ and $A_i \neq \neg B_j$ for $i = 1, \ldots, n$ and $j = 1, \ldots, k$. If $X \Vdash\!\!\!\prec Y$, then the ordered sequent:*

$$A_1^{[1]}, \ldots, A_n^{[n]}, \neg B_1^{[n+1]}, \ldots, \neg B_k^{[n+k]} \vdash$$

is provable in $\mathsf{MI^{CPL}}$*.*

Proof. By Theorem 5 and Theorem 13. □

As for strong sc-entailment, one gets analogous results by applying Theorem 8 instead of Theorem 5.

7 Some Conceptual Applications

7.1 Deep Contraction

Let us imagine that we are working with a consistent non-empty set of CPL-wffs X (for instance, representing a database or a belief base) and that a contingent CPL-wff B has been derived from X. Assume that the derivation mechanism used preserves CPL-entailment. Now suppose that we have strong, though independent from X, reasons to believe that $\neg B$ rather than B is the case. As long as we stick to Classical Logic, extending X with $\neg B$ is not a good move. An option is to switch to some non-monotonic logic and its consequence operation. As we have shown, strong sc-entailment is not monotone. But no extension of X produces $\neg B$ as a strongly sc-entailed consequence of X. This is due to:

Corollary 38. *If $X \models\!\!\!\prec B$, then there is no proper superset Z of X such that $Z \models\!\!\!\prec \neg B$.*

Proof. Let $X \models\!\!\!\prec B$. Suppose that $Z \models\!\!\!\prec \neg B$, where $X \subseteq Z$ and $X \neq Z$. If $X \models\!\!\!\prec B$, then $X \models B$ and hence $Z \models B$. If $Z \models\!\!\!\prec \neg B$, then $Z \models \neg B$. Therefore Z is inconsistent and thus, by Corollary 14, $Z \not\models\!\!\!\prec \neg B$. □

A rational move is to contract X first, and in a way that prevents the appearance of B as a conclusion of any legitimate (i.e. preserving classical entailment) derivation from the contracted set. How to achieve this? One can examine the derivation of B from X that has actually been performed, identify the elements of X used as premises, and then contract X by removing from it at least one wff which was used as a premise in the performed derivation. This, however, will not do: it is possible that B is classically entailed by many subsets of X, including some that do not contain the just removed wff(s), and thus B can still be legitimately derived from the set contracted in the above manner. Examining all possible legitimate derivations of B from X constitutes a difficult if not a hopeless task. However, a solution is suggested by the content of Theorem 11. By and large, it suffices to consider all the finite subsets of X that strongly sc-entail B, and to remove from X exactly one element of every such subset. A contracted set obtained in this way does not CPL-entail the wff B and therefore no legitimate derivation leads from the set to B.

Remark 9. The way of proceeding proposed above is akin to (but not identical with) the well-known idea of consistency restoring by calculating a minimal hitting set of the family of all minimally inconsistent subsets of an inconsistent set in order to eliminate elements of the hitting set from the inconsistent set in question.[21] The general idea goes back to [15] and gave rise to some related constructions.[22] However, contraction of the analysed kind does not aim at consistency restoring, but at making a legitimate deduction of B from the resultant set impossible. These are interconnected, but yet different issues.

In what follows we apply some conceptual tools taken from [29].

Let $\mathbb{F}$ be a family of sets, i.e. a set of sets. These sets need not be disjoint. In the first step we define a related family of pairwise disjoints sets.

Definition 9. $\mathbb{F}^{\otimes} =_{df} \{X^{\otimes} : X \in \mathbb{F}\}$, where:

$$X^{\otimes} = \begin{cases} X \times \{X\} & \text{if } X \neq \emptyset, \\ \emptyset & \text{if } X = \emptyset. \end{cases}$$

Since the elements of $\mathbb{F}^{\otimes}$ are pairwise disjoint, by the Axiom of Choice we get:

---

[21] A set X is a hitting set of a family of sets $\mathbb{F}$ iff $X \cap Y \neq \emptyset$ for each $Y \in \mathbb{F}$. A hitting set of $\mathbb{F}$ is minimal if no proper subset of it is a hitting set of $\mathbb{F}$. Hitting sets are also called choice sets. For hitting/choice sets see, e.g., [24], pp. 335–338.

[22] Cf. e.g., [3].

Corollary 39. *If $\mathbb{F}^\otimes \neq \emptyset$ and $\emptyset \notin \mathbb{F}^\otimes$, then there exists a set γ such that γ comprises exactly one element, $\langle A, X \rangle$, of each $X^\otimes \in \mathbb{F}^\otimes$.*

We introduce an auxiliary notion.

Definition 10. *γ is a $\chi^\otimes(\mathbb{F})$-set iff*

1. *$\gamma \subseteq \bigcup \mathbb{F}^\otimes$ and*

2. *for each $X^\otimes \in \mathbb{F}^\otimes$ such that $X^\otimes \neq \emptyset$ there exists exactly one $\langle A, X \rangle \in X^\otimes$ such that $\langle A, X \rangle \in \gamma$.*

A $\chi^\otimes(\mathbb{F})$-set is a set of ordered pairs. We take into account the first projection of the set.

Definition 11. *Let γ be a $\chi^\otimes(\mathbb{F})$-set.*

$$\gamma^1 =_{df} \{A : \langle A, X \rangle \in \gamma\}.$$

Now we are able to introduce the crucial technical notion.

Definition 12. *Z is a $\chi(\mathbb{F})$-set iff $Z = \gamma^1$ for some $\chi^\otimes(\mathbb{F})$-set γ.*

By and large, a $\chi(\mathbb{F})$-set is a set comprising exactly one *representative*, with regard to the above construction, of each non-empty set belonging to a family of sets $\mathbb{F}$. The representatives of distinct sets in a χ-set need not be distinct. One should not confuse the existence of exactly one representative (of the above kind) of each set belonging to a family of sets with the existence of a system of distinct representatives of the family.[23] One can prove that a χ-set always exists, i.e. for any family of sets $\mathbb{F}$ there exists a $\chi(\mathbb{F})$-set (cf. [29]).

Let us now come back to the contraction issue. The following holds.

Theorem 18 (Deep contraction). *Let X be a consistent non-empty set of wffs, and let B be a non-valid wff such that $X \models B$. Let $\mathbb{F} = \{W \subseteq X : W \not\models B\}$, and let Z be a $\chi(\mathbb{F})$-set. Then $(X \setminus Z) \not\models B$.*

Proof. By Theorem 11, the family $\mathbb{F}$ is non-empty. If B is non-valid, $\emptyset \notin \mathbb{F}$. Thus $X' \neq \emptyset$ for each $X' \in \mathbb{F}$, and hence $Z \neq \emptyset$.

The set $X \setminus Z$ is consistent, since X is, by assumption, consistent.

Suppose that $(X \setminus Z) \models B$. It follows that $(X \setminus Z) \neq \emptyset$ (as B is not valid) and, by Theorem 11, that $Y \not\models B$ for some finite subset Y of $X \setminus Z$. Moreover, $Y \neq \emptyset$;

---

[23] As it is well-known, a system of distinct representatives – a transversal of a family of sets – does not always exist; cf., e.g., [26], Chapter 8.

otherwise B would have been valid. But the only subsets of X that strongly sc-entail B are the sets in $\mathbb{F}$. Hence $Y = X°$ for some element, $X°$, of $\mathbb{F}$. But $(X' \cap Z) \neq \emptyset$ for each $X' \in \mathbb{F}$. Hence $(X° \cap Z) \neq \emptyset$. On the other hand, $(Y \cap Z) = \emptyset$ due to the fact that Y is a subset of $X \setminus Z$. It follows that $Y \neq X°$. We arrive at a contradiction. Therefore $(X \setminus Z) \not\models B$. □

Let us stress that Theorem 18 speaks about *any* χ-set of the family of subsets of X which strongly sc-entail B. There are usually many such sets. Each of them may be subtracted from X in order to arrive at a subset of X that does not (classically) entail B. In other words, "deep contraction" can be successfully performed in many ways and its outcome depends on the χ-set chosen.[24] As for the multiplicity of possible outcomes, and their dependence on factors different from the set subjected to be contracted and the wff w.r.t. which the operation is performed, deep contraction does not differ from other contraction operations characterized in belief revision theories. Note, however, that deep contraction has a kind of computational flavour. In order to perform it one needs a χ-set of the family of subsets of X which strongly sc-entail B, and this requires that the family has to be "calculated" first. Given the content of Theorem 5, this, in turn, can be achieved by identifying all the minimally inconsistent subsets of an inconsistent set of some kind.[25] Algorithms for solving such problems are already known in the literature.[26]

Remark 10. A set of wffs X supposed to be contracted w.r.t. B may be either finite or infinite. In the latter case it can happen that the family of subsets of X that sc-entail B is countably or even uncountably infinite. It follows that the relevant χ-sets may be infinite. However, we are dealing here with Classical Logic, in which entailment is compact: everything entailed by an infinite set of wffs is also entailed by some finite subset(s) of the set. One can easily prove:

---

[24] A simple example may be of help. Let $X = \{p \vee q \to r, p, q\}$ and $B = r$. The relevant family of MI-sets comprises $\{p \vee q \to r\}$ and $\{p \vee q \to r, q\}$; let us designate it by $\mathbb{F}$. The $\chi(\mathbb{F})$-sets are: (1) $\{p, q\}$, (2) $\{p \vee q \to r\}$, (3) $\{p \vee q \to r, q\}$, (4) $\{p \vee q \to r, p\}$. The result of deep contraction of X, depending on the $\chi(\mathbb{F})$-set used, is (1') $\{p \vee q \to r\}$, or (2') $\{p, q\}$, or (3') $\{p\}$, or (4') $\{q\}$. Which $\chi(\mathbb{F})$-set is to be used depends on epistemic factors. By the way, the example presented above shows that deep contraction does not amount to subtracting a minimal choice set of the family of all MI-sets in question.

Belief revision theories view contraction as an operation which is supposed to achieve its goal(s) in an "economical" manner: the loss should be kept to a minimum. This means many things, depending on an account advocated. As for deep contraction, the "minimalization of loss" issue is only of a secondary importance.

[25] More specifically, all minimally inconsistent subsets of $X \cup \{\neg B\}$ such that $\neg B$ belongs to each of them have to be identified first. Then the family $\{Y \subseteq X : Y \cup \{\neg B\}$ is a MI-set and '$\neg B$' $\notin Y\}$ constitutes the solution.

[26] See, e.g., [12], and [3].

Corollary 40. *Let X be an infinite consistent set of wffs, and let B be a non-valid wff such that $X \models B$. If Y is a finite subset of X such that $Y \models B$, and Z is a $\chi(\mathbb{F}^*)$-set, where $\mathbb{F}^* = \{W \subseteq Y : W \not\Kappa B\}$, then $X \cap (Y \setminus Z) \not\models B$.*

Proof. Suppose otherwise. Then $(Y \setminus Z) \models B$, contrary to Theorem 18. □

Thus when entailment is compact, an infinite set X can also be "deeply contracted" w.r.t. B without relying on infinite χ-set(s) that correspond(s), in the way described above, to the whole X. It suffices to use a χ-set which corresponds to a finite subset Y of X that classically entails B. Needless to say, the resultant set $X \cap (Y \setminus Z)$ will be finite.

7.2 Argument Analysis

7.2.1 Strong Entailments and Lehrer's Notion of Relevant Deductive Argument

Strong sc-entailment is a special case of strong mc-entailment. There are, however, close affinities between the concept of strong sc-entailment and the notion of relevant deductive argument introduced long ago by Keith Lehrer (cf. [9]). Here is Lehrer's definition:

> An argument RD is a relevant deductive argument if and only if RD contains a non-empty set of premises $P_1, P_2, \ldots, P_n$ and a conclusion C such that a set of statements consisting of just $P_1, P_2, \ldots, P_n$, and $\neg C$ (or any truth functional equivalent of $\neg C$) is a minimally inconsistent set. A set of statements is a minimally inconsistent set if and only if the set of statements is logically inconsistent and such that every proper subset of the set is logically consistent. ([9], p. 298.)

Is strong sc-entailment just the semantic relation that holds between premises and conclusions of Lehrer's relevant deductive arguments? As Theorem 8 illustrates, the fact that $\{P_1, \ldots, P_n, \neg C\}$ is a MI-set is a necessary but insufficient condition for $\{P_1, P_2, \ldots, P_n\} \Kappa C$ to hold. It is additionally required that '$\neg C$' $\notin \{P_1, P_2, \ldots, P_n\}$. On the other hand, the statement "a set of statements consisting of just $P_1, P_2, \ldots, P_n$, and $\neg C$ (or any truth functional equivalent of $\neg C$)" seems to secure that the additional requirement is to be met. Anyway, relevant deductive arguments in Lehrer's sense (henceforth: *rd-arguments*) and deductive arguments in which the conclusion is *strongly* sc-entailed by the premises – let us refer to them as to $\Kappa$-arguments – share basic properties. In both cases, as stipulated by Lehrer and witnessed by corollaries 22 and 16, only contingent wffs can serve as premises and conclusions, and the set of premises is always consistent. Assuming that a set

of premises of an argument must be non-empty, there is neither rd-argument nor
⊨-argument which leads to a valid wff or to a contradictory/inconsistent conclusion
(cf. corollaries 20 and 17, respectively). So some classes of relevant (in the intuitive
sense of the word) deductive arguments are beyond the scope of either analysis.
On the other hand, our considerations, though indirectly, throw new light of the
properties of rd-arguments. Theorem 11 yields that for each deductive argument $\mathcal{A}$
from a consistent set of premises there exists a corresponding ⊨-argument leading
from a finite subset of the set of premises of $\mathcal{A}$ to the conclusion of the argument
$\mathcal{A}$. The premises of an ⊨-argument are mutually independent (cf. Corollary 25),
and their conclusions always share variable(s) with the premises (cf. Corollary 24).
A multi-premise ⊨-argument differs from the respective single-premise ⊨-argument
based on a conjunction of premises of the multi-premise argument (cf. section 4.2.3).

It seems that the intuitive concept of *linked* multi-premise deductive argument
can be successfully explicated in terms of strong sc-entailment: a linked multi-
premise deductive argument is an argument whose conclusion is strongly sc-entailed
by the set of premises.

7.2.2 Multiple-Conclusion Arguments?

The concept of argument is sometimes generalized to include arguments containing a finite number of conclusions. As a result, one gets an unproblematic class of
arguments having exactly one conclusion – let us call them *sc-arguments* – and a
problematic class of arguments having at least two (but still finitely many) conclusions. Let us call the latter *mc-arguments*.

As we observed, strong sc-entailment is, in principle, the semantic relation which
holds between premises and conclusions of relevant (in the Lehrer's sense) deductive
sc-arguments. By analogy, strong mc-entailment can be viewed as singling out an
important class of mc-arguments. We coin them (for the lack of a better idea)
germane mc-arguments. To be more precise, by a germane mc-argument we mean
an mc-argument whose premises strongly mc-entail the *set* of conclusions of the
argument. The properties of strong mc-entailment pointed out above seem to speak
in favour of this proposal.

There is an ongoing discussion as to whether mc-arguments are artifacts (cf. [17],
[23], [2]). An mc-argument is often identified with the corresponding sc-argument
whose conclusion is a disjunction of all the "conclusions" of the respective mc-
argument. Without pretending to resolve the issue, let us only notice the following.

Consider:

$$\frac{p}{p \vee q} \tag{40}$$

and
$$\frac{p}{p,q} \quad (41)$$

Since $p \mathrel{\kappa} p \vee q$ holds, (40) constitutes an $\mathrel{\kappa}$-argument. But $p \mathrel{\|\kappa} \{p,q\}$ does not hold and thus (41) is not a germane mc-argument. So it happens that, having premises fixed, there exist $\mathrel{\kappa}$-arguments leading from the premises(s) to a disjunction, but there is no germane mc-argument that leads from the premise(s) to the set of disjuncts.

Now let us consider:
$$\frac{p \vee q}{p,q} \quad (42)$$

and
$$\frac{p \vee q}{p} \quad (43)$$

$$\frac{p \vee q}{q} \quad (44)$$

(42) is a germane mc-argument, while (43) and (44) are not $\mathrel{\kappa}$-arguments. This is not an exception, but a rule. Due to Corollary 2, if an mc-argument leading from a disjunction to the set of disjuncts is germane, then there is no $\mathrel{\kappa}$-argument that leads from the disjunction to a single disjunct. In general, the existence of a germane mc-argument from X to Y excludes the existence of an $\mathrel{\kappa}$-argument from X to a single conclusion that belongs to Y. Similarly, the existence of an $\mathrel{\kappa}$-argument leading from X to a wff which is only one of the elements of a set of wffs Y suppresses the existence of a germane mc-argument leading from X to Y.

8 Final Remarks

8.1 The First-Order Case

So far we have dealt with the classical propositional case. So a natural question arises: what, if anything, will change when we move to the first-order level and consider strong entailments based on First-Order Logic (**FOL**)?

As it is well-known, sc-entailment in **FOL** can be defined either in terms of satisfaction or in terms of truth, and similarly for mc-entailment. However, truth of a wff in a **FOL**-model equals satisfaction of the wff *under all assignments* of values to individual variables, where the values belong to the universe of the model. Therefore the respective concepts of entailment do not coincide when sentential functions, that is, wffs in which free variables occur, enter the picture, although they coincide on **FOL**-sentences (i.e. wffs with no free variables). Similarly, inconsistency can be defined either as unsatisfiability or as the lack of a FOL-model which makes true all the wffs in question. These are not the same thing if sentential functions are allowed.[27]

When one wants to move from the propositional level to the first-order one, three possibilities emerge.

The simplest solution is to assume that strong entailments, as well as the other semantic notions employed, are defined for sentences only. The concept of truth under a CPL-valuation is to be replaced with the concept of truth in a FOL-model. Then the results concerning CPL "translate" into the respective results concerning the "sentential part" of FOL. Of course, this does not pertain to results which rely on the assumption that the wffs considered are propositional, in particular to Theorem 4. Needless to say, an analogous remark applies to the other options presented below.

The second option is to allow for sentential functions and to replace "true under a CPL-valuation" with "satisfied in a FOL-model under an assignment of values to individual variables." In such a case inconsistency would mean unsatisfiability. There is, however, a price to be paid. While sc-entailment defined in terms of satisfaction ensures the transmission of truth, mc-entailment defined by means of satisfaction (i.e. roughly, by the clause: "for every assignment ι: if all the wffs in X are satisfied under ι, then at least one wff in Y is satisfied under ι") does not warrant the existence of a *true* wff in Y when all the wffs in X are true. This lack of warranty shows up in the case of mc-entailed sets containing sentential functions. As a consequence, the intuitive meaning of the concept of strong mc-entailment changes.

As for the third option, one allows for sentential functions and replaces "true under a CPL-valuation" with "true in a FOL-model." Now consistency of a set of wffs would mean the existence of a FOL-model which makes all the wffs true. Contingent wffs are these which are true in some, but not all FOL-models. However, sc-entailment of A from X amounts to inconsistency of the set comprising X and *the negation of the universal closure* of A. Similarly, mc-entailment between X and Y holds iff the set $X, \neg \overline{Y}$ is inconsistent, where $\overline{Y}$ is the set of universal closures of elements of Y. So a "translation" of results concerning CPL should be performed with caution. In particular, whenever consistency/inconsistency of propositional formulas of the form $\neg A$ or sets of such formulas have been considered, first-order wffs of the form $\neg \overline{A}$, where $\overline{A}$ is the universal closure of A, should be used. For example, the FOL counterparts of theorems 5 and 8 now are:

$$X \parallel\!\sim Y \quad \text{iff} \quad X \cap \neg \overline{Y} = \emptyset \text{ and } X, \neg \overline{Y} \text{ is a MI-set.}$$

---

[27]For instance, the set $\{P(x), \neg \forall x P(x)\}$, where P is a one-place predicate, is satisfiable, but there is no FOL-model which makes its elements simultaneously true.

$$X \mathrel{\mspace{1mu}\vdash\mspace{-8mu}\mathrel{\triangleleft}} B \quad \text{iff} \quad \text{`}\ulcorner\neg B\urcorner\text{'} \notin X \text{ and } X, \neg B \text{ is a MI-set}.$$

Another example is this. What we have called "deduction theorems" for strong entailments (cf. theorems 7 and 26), relied upon Corollary 11. However, its counterpart does not hold for FOL when entailments are defined in terms of truth. Instead, we have:

$$Z, \overline{A} \models W \quad \text{iff} \quad Z \models \ulcorner \overline{A} \to W \urcorner.$$

As a consequence, in order to get counterparts of theorems 7 and 26 one has to replace A with $\overline{A}$. An analogous remark pertains to corollaries 12, 27, and 28.

8.2 Further Research: Strong Entailments in Non-Classical Logics

In this paper we have concentrated upon Classical Logic. A natural next step is to turn to non-classical logics. Which of the results presented above would remain valid if we defined strong entailments in terms of entailments based on a non-classical logic? No doubt, this is an interesting question. Yet, it deserves a separate paper or even a series of papers. So let me only comment on the relation between the concepts of strong entailments and the concept of minimally inconsistent set. Theorems 5 and 8 (as well as their counterparts for FOL) show how these concepts are interconnected for Classical Logic. However, analogues of theorems 5 and 8 fail in some non-classical logics. Negationless logics provide trivial examples here, but there are others. For instance, in Intuitionistic Logic (INT) the following:

$$\{\neg\neg p, \neg p\} \models_{\mathsf{INT}} \bot$$

$$\{\neg\neg p\} \not\models_{\mathsf{INT}} \bot$$

$$\{\neg p\} \not\models_{\mathsf{INT}} \bot$$

hold and thus $\{\neg\neg p, \neg p\}$ can be regarded as a MI-set. Needless to say, '$\neg p$' $\notin \{\neg\neg p\}$. On the other hand, we have:

$$\neg\neg p \not\models_{\mathsf{INT}} p$$

and hence, assuming that strong sc-entailment presupposes sc-entailment, $\neg\neg p$ and p are not linked with strong sc-entailment. It follows that the "intuitionistic" analogues of theorems 8 and 5 do not hold.

Appendix: Soundness and Completeness of MI^{CPL}

In order to prove soundness and completeness of the calculus MI^{CPL} with respect to MI-sets we need a series of corollaries and lemmas.

As an immediate consequence of definitions introduced in Section 6 we get

Corollary 41. *X is a MI-set iff there exists an ordered sequent $C_{\vec{m}}^{[\vec{m}]} \vdash$ such that $X = \mathsf{wff}(C_{\vec{m}}^{[\vec{m}]} \vdash)$ and*

1. *$\mathsf{wff}(C_{\vec{m}}^{[\vec{m}]} \vdash)$ is an inconsistent set, and*

2. *for each $k \in \{1, \ldots, m\}$: the set $\mathsf{wff}(f_{\backslash [k]}(C_{\vec{m}}^{[\vec{m}]}) \vdash)$ is consistent.*

The following hold:

Lemma 2.

1. *$\mathsf{wff}(S \,'\, \alpha^{[i]} \,'\, T \vdash)$ is inconsistent iff $\mathsf{wff}(S \,'\, \alpha_1^{[i]} \,'\, \alpha_2^{[i]} \,'\, T \vdash)$ is inconsistent.*

2. *$\mathsf{wff}(S \,'\, \beta^{[i]} \,'\, T \vdash)$ is inconsistent iff $\mathsf{wff}(S \,'\, \beta_1^{[i]} \,'\, T \vdash)$ is inconsistent and $\mathsf{wff}(S \,'\, \beta_2^{[i]} \,'\, T \vdash)$ is inconsistent.*

3. *$\mathsf{wff}(S \,'\, A^{[i]} \,'\, T \vdash)$ is inconsistent iff $\mathsf{wff}(S \,'\, \neg\neg A^{[i]} \,'\, T \vdash)$ is inconsistent.*

Lemma 3.

1. *If $\Phi; \sigma \vdash; \Psi$ results from $\Phi; \theta \vdash; \Psi$ by a rule of MI^{CPL}, then the set $\mathsf{wff}(\sigma \vdash)$ is inconsistent iff $\mathsf{wff}(\theta \vdash)$ is an inconsistent set.*

2. *If $\Phi; \sigma_1 \vdash; \sigma_2 \vdash; \Theta$ results from $\Phi; \theta \vdash; \Psi$ by a rule of MI^{CPL}, then both $\mathsf{wff}(\sigma_1 \vdash)$ and $\mathsf{wff}(\sigma_2 \vdash)$ are inconsistent sets iff $\mathsf{wff}(\theta \vdash)$ is an inconsistent set.*

Proof. If $\Phi; \sigma \vdash; \Psi$ results from $\Phi; \theta \vdash; \Psi$ by a rule of MI^{CPL}, then θ involves a numerically annotated α-wff or a numerically annotated double negated wff. But $X, \alpha \models \emptyset$ iff $X, \alpha_1, \alpha_2 \models \emptyset$, and $X, \neg\neg A \models \emptyset$ iff $X, A \models \emptyset$.

If $\Phi; \sigma_1 \vdash; \sigma_2 \vdash; \Theta$ results from $\Phi; \theta \vdash; \Psi$ by a rule of MI^{CPL}, then a numerically annotated β-wff is a term of θ. Yet, $X, \beta \models \emptyset$ iff $X, \beta_1 \models \emptyset$ and $X, \beta_2 \models \emptyset$. □

Lemma 4. *If a seqsequent Ψ results from a seqsequent Φ by a rule of MI^{CPL}, then the following conditions are equivalent:*

1. *for each constituent $\sigma \vdash$ of Φ: the set $\mathsf{wff}(\sigma \vdash)$ is inconsistent,*

2. *for each constituent $\theta \vdash$ of Ψ: the set $\mathsf{wff}(\theta \vdash)$ is inconsistent.*

Proof. By Lemma 3. □

Theorem 12 (Soundness w.r.t. MI-sets). *Let X be a finite non-empty set of wffs, and let $\sigma \vdash$ be an ordered sequent such that $\text{wff}(\sigma \vdash) = X$. If the sequent $\sigma \vdash$ is provable in MI^{CPL}, then X is a MI-set.*

Proof. Let $C_{\vec{m}}^{[\vec{m}]} \vdash$ be an arbitrary but fixed ordered sequent such that $X = \text{wff}(C_{\vec{m}}^{[\vec{m}]} \vdash)$. Assume that

$$\Theta_1, \ldots, \Theta_n \tag{45}$$

is a proof of the sequent $C_{\vec{m}}^{[\vec{m}]} \vdash$ in MI^{CPL}. By Definition 7, each constituent of Θ_n is a closed atomic sequent. Hence the set $\text{wff}(\theta \vdash)$ is inconsistent for each constituent $\theta \vdash$ of Θ_n. Therefore, by Lemma 4, the set $\text{wff}(C_{\vec{m}}^{[\vec{m}]} \vdash)$ is inconsistent, that is, X is inconsistent.

We shall prove the following:

(★) *if Θ_{j+1} has a constituent, $\sigma \vdash$, such that the set $\text{wff}(f_{\backslash [k]}(\sigma) \vdash)$ is consistent, then Θ_j has a constituent, θ, such that the set $\text{wff}(f_{\backslash [k]}(\theta) \vdash)$ is consistent, where $1 \leqslant j < n$ and $1 \leqslant k \leqslant m$.*

Let $\sigma \vdash$ be a constituent of Θ_{j+1} for which the set $\text{wff}(f_{\backslash [k]}(\sigma) \vdash)$ is consistent. Recall that rules of MI^{CPL} "act locally": if a rule is applied to a seqsequent, only one constituent and only one occurrence of a na-wff in the constituent are acted upon (more precisely, only one term of the sequence of na-wffs which occurs in the constituent is transformed). When Θ_{j+1} results from Θ_j by a rule, the following cases are possible:

(a) $\sigma \vdash$ has been rewritten from Θ_j into Θ_{j+1} (since a rule has been applied to Θ_j w.r.t. some other constituent of it),

(b) the occurrence of $\sigma \vdash$ in Θ_{j+1} is due to an application of a rule to Θ_j w.r.t. a constituent, say, $\theta \vdash$, of Θ_j.

If (a) is the case, then (★) holds trivially. So assume that (b) holds. Two sub-cases are possible:

(b_1) a rule has been applied to Θ_j w.r.t. the constituent $\theta \vdash$ and a term of θ annotated with k,

(b_2) a rule has been applied to Θ_j w.r.t. the constituent $\theta \vdash$ and a term of θ which is annotated with some numeral j different from k.

If (b_1) holds, then $\mathsf{wff}(f_{\backslash[k]}(\theta) \vdash) = \mathsf{wff}(f_{\backslash[k]}(\sigma) \vdash)$, so the set $\mathsf{wff}(f_{\backslash[k]}(\theta) \vdash)$ is consistent. Assume that (b_2) is the case. Suppose that the set $\mathsf{wff}(f_{\backslash[k]}(\theta) \vdash)$ is inconsistent though $\mathsf{wff}(f_{\backslash[k]}(\sigma) \vdash)$ is a consistent set. Both $\mathsf{wff}(f_{\backslash[k]}(\sigma) \vdash)$ and $\mathsf{wff}(f_{\backslash[k]}(\theta) \vdash)$ do not contain wffs annotated with k. So the hypothetical inconsistency of the set $\mathsf{wff}(f_{\backslash[k]}(\theta) \vdash)$ is due to the occurrence in θ of some wffs(s) annotated with numeral(s) different from k. Observe that the inconsistency of $\mathsf{wff}(f_{\backslash[k]}(\theta) \vdash)$ yields the inconsistency of the set $\mathsf{wff}(\theta \vdash)$. However, $\sigma \vdash$ is a constituent of Θ_{j+1} because a rule has been applied to Θ_j w.r.t. $\theta \vdash$ and a wff annotated with a numeral different from k. Thus, by Lemma 2, the set $\mathsf{wff}(\sigma \vdash)$ is inconsistent. Moreover, its inconsistency is due to the occurrence in σ of wffs annotated with numerals different from k. Therefore the set $\mathsf{wff}(f_{\backslash[k]}(\sigma) \vdash)$ is inconsistent. We arrive at a contradiction. This completes the proof of ($\bigstar$).

The sequence (45) is supposed to be a proof, so, by Definition 7, for any $k \in \{1, \ldots, m\}$ there exists a constituent, say, $\rho \vdash$, of Θ_n such that, as $f_{\backslash[k]}(\rho \vdash)$ is either $\emptyset \vdash$ or is an open atomic sequent, the set $\mathsf{wff}(f_{\backslash[k]}(\rho) \vdash)$ is consistent. Thus, by ($\bigstar$) proven above, any term/seqsequent of (45) has a constituent, $\zeta \vdash$, such that $\mathsf{wff}(f_{\backslash[k]}(\zeta) \vdash)$ is a consistent set of wffs. But the sequent $C_{\overrightarrow{m}}^{[\overrightarrow{m}]} \vdash$ is the only constituent of Θ_1. Hence $\mathsf{wff}(f_{\backslash[k]}(C_{\overrightarrow{m}}^{[\overrightarrow{m}]}) \vdash)$ is a consistent set. As k was an arbitrary element of $\{1, \ldots, m\}$, by Corollary 41 it follows that X is a MI-set. □

An auxiliary concept is needed.

Definition 13 (MI^{CPL}-transformation of a sequent). *A MI^{CPL}-transformation of a sequent $\sigma \vdash$ is a finite sequence $\Theta_1, \ldots, \Theta_n$ of seqsequents such that: (a) $\Theta_1 = \langle \sigma \vdash \rangle$, and (b) Θ_{j+1} results from Θ_j by a rule of MI^{CPL} for $1 \leqslant j < n$.*

Theorem 13 (Completeness w.r.t MI-sets). *If X is a MI-set, then any ordered sequent $\sigma \vdash$ such that $\mathsf{wff}(\sigma \vdash) = X$ is provable in MI^{CPL}.*

Proof. A moment's reflection on the rules of MI^{CPL} reveals that for each ordered sequent $\sigma \vdash$ such that $\mathsf{wff}(\sigma \vdash)$ is an inconsistent set of wffs, there exist MI^{CPL}-transformations of the sequent which end with seqsequents whose constituents are closed atomic sequents only.

Assume that X is a MI-set, and that $C_{\overrightarrow{m}}^{[\overrightarrow{m}]} \vdash$ is an ordered sequent such that $\mathsf{wff}(C_{\overrightarrow{m}}^{[\overrightarrow{m}]} \vdash) = X$. Since X is a MI-set, $\mathsf{wff}(C_{\overrightarrow{m}}^{[\overrightarrow{m}]} \vdash)$ is an inconsistent set. Let

$$\Theta_1, \ldots, \Theta_n \qquad (46)$$

be a MI^{CPL}-transformation of the sequent $C_{\overrightarrow{m}}^{[\overrightarrow{m}]} \vdash$ such that each constituent of Θ_n is a closed atomic sequent. Suppose that the transformation (46) is not a proof of $C_{\overrightarrow{m}}^{[\overrightarrow{m}]} \vdash$.

The transformation (46) can be depicted as:

$$\Theta_1 \leftarrow\!\!\hookleftarrow \mathsf{R}_x^{[i_1]} \qquad (47)$$

$$\Theta_2 \leftarrow\!\!\hookleftarrow \mathsf{R}_x^{[i_2]}$$

$$\ldots$$

$$\Theta_{n-1} \leftarrow\!\!\hookleftarrow \mathsf{R}_x^{[i_{n-1}]}$$

$$\Theta_n$$

where '$\leftarrow\!\!\hookleftarrow \mathsf{R}_x^{[i_j]}$' indicates that the rule applied to Θ_j acts upon a wff annotated with i_j (more precisely, upon an occurrence of such a wff in a sequent that belongs to Θ_j).

If the transformation (46) is not a proof, then, by Definition 7, there exists an index k, where $1 \leqslant k \leqslant m$, such that for each sequent $\theta \vdash$ which occurs in Θ_n, $f_{\setminus [k]}(\theta) \vdash$ is a closed atomic sequent.

Suppose that $m = 1$. Thus X is a singleton set, each rule of (47) acts upon a wff annotated with 1, and all the wffs which occur in Θ_n are annotated with 1. Hence $f_{\setminus [1]}(\theta) \vdash = \emptyset \vdash$ for any constituent $\theta \vdash$ of Θ_n. It follows that there is no constituent of Θ_n such that $f_{\setminus [1]}(\theta \vdash)$ is a closed atomic sequent. We arrive at a contradiction. Thus $m \neq 1$.

We proceed as follows. First, we remove from (47) each Θ_j which is associated with $\leftarrow\!\!\hookleftarrow \mathsf{R}_x^{[k]}$, that is, we skip all the lines of (47) in which a rule acts upon a wff annotated with k. Let

$$\Theta_1^*, \ldots, \Theta_h^* \qquad (48)$$

stand for the subsequence of (46) obtained from it in this way. Each Θ_j^*, where $1 \leqslant j \leqslant h$, is a sequence of sequents. Let $\Theta_j^* = \langle \xi_1 \vdash, \ldots, \xi_s \vdash \rangle$. We define Θ_j^{**} as:

$$\langle f_{\setminus [k]}(\xi_1) \vdash, \ldots, f_{\setminus [k]}(\xi_s) \vdash \rangle \qquad (49)$$

Since $f_{\setminus [k]}(\xi_i) \vdash$ is a sequent for $1 \leqslant i \leqslant s$, (49) is a seqsequent. Then we consider the following sequence of seqsequents:

$$\Theta_1^{**}, \ldots, \Theta_h^{**} \qquad (50)$$

Observe that wffs annotated with k do not occur in any constituent of any element/term of (50). Clearly, $f_{\setminus [k]}(C_{\overrightarrow{m}}^{[\overrightarrow{m}]}) \vdash$ is

$$C_1^{[1]}, \ldots, C_{k-1}^{[k-1]}, C_{k+1}^{[k+1]}, \ldots, C_m^{[m]} \vdash \qquad (51)$$

It is easily seen that (50) is a MI^{CPL}-transformation of the sequent (51). On the other hand, each constituent of Θ_h^{**} is a closed atomic sequent and hence wff$(\theta \vdash)$ is inconsistent for any constituent $\theta \vdash$ of Θ_h^{**}. Thus, by Lemma 4,

$$\{C_1, \ldots, C_{k-1}, C_{k+1}, \ldots, C_m\} \tag{52}$$

is an inconsistent set of wffs. But (52) is a proper subset of X. Therefore X is not a MI-set. We arrive at a contradiction. This completes the proof. □

Theorem 14. *If there exists a MI^{CPL}-disproof of an ordered sequent $C_{\overrightarrow{m}}^{[\overrightarrow{m}]} \vdash$, then the set wff$(C_{\overrightarrow{m}}^{[\overrightarrow{m}]} \vdash)$ is inconsistent, but is not a MI-set.*

Proof. Let

$$\Theta_1', \ldots, \Theta_n' \tag{53}$$

be an arbitrary but fixed disproof of $C_{\overrightarrow{m}}^{[\overrightarrow{m}]} \vdash$. Everything what has been said, in the above proof of Theorem 13, about the transformation (46), can be repeated with regard to the disproof (53) (of course, after replacing Θ_j with Θ_j' for $1 \leqslant j \leqslant n$). So wff$(C_{\overrightarrow{m}}^{[\overrightarrow{m}]} \vdash)$ is not a MI-set. Yet, due to Definition 8 and Lemma 3, it is an inconsistent set. □

Theorem 15. *If X is a finite inconsistent set of wffs which is not a MI-set, then any ordered sequent $\sigma \vdash$ such that wff$(\sigma \vdash) = X$ is disprovable in MI^{CPL}.*

Proof. Let X be an arbitrary but fixed finite inconsistent set of wffs which is not a MI-set. We define:

$$\Sigma = \{\sigma \vdash : \mathsf{wff}(\sigma \vdash) = X \text{ and } \sigma \vdash \text{ is an ordered sequent}\}$$

Let Λ be the set of all MI^{CPL}-transformations of sequents in Σ. As X is inconsistent, Λ includes a non-empty subset Λ_0 of MI^{CPL}-transformations each of which ends with a seqsequent involving closed atomic sequent(s) only. By assumption, X is not a MI-set. Thus, by Theorem 12, no element of Λ_0 has a MI^{CPL}-proof. Therefore each transformation in Λ_0 violates the fourth clause of Definition 7. Hence Λ_0 comprises MI^{CPL}-disproofs of the sequents in Σ. □

References

[1] A. Avron. The method of hypersequents in the proof theory of propositional non-classical logics. In W. Hodges, M. Hyland, C. Steinhorn and J. Truss, editors, *Logic: Foundations to applications*, pages 1–32. Oxford Science Publications, Oxford, 1996.

[2] J. C. Beall. Multiple-conclusion LP and default classicality. *The Review of Symbolic Logic*, 4:326–336, 2011.

[3] G. Brewka, M. Thimm and M. Ulbricht. Strong Inconsistency in Nonmonotonic Reasoning. *Proceedings of the Twenty-Sixth International Joint Conference on Artificial Intelligence, IJCAI-17*, pages 901–907, 2017.

[4] R. Carnap. *Formalization of Logic*. Harvard University Press, Cambridge, MA, 1943.

[5] G. Gentzen. Investigation into logical deduction. In M. E. Szabo, editor, *The Collected Papers of Gerhard Gentzen*, pages 68–131. North-Holland, 1969.

[6] A. Grzegorczyk. *An Outline of Mathematical Logic*. D. Reidel Publishing Company, Dordrecht, 1974.

[7] A. Grzelak and D. Leszczyńska-Jasion. Automatic proof generation in axiomatic system for CPL by means of the method of Socratic proofs. *Logic Journal of the IGPL*, 28: 109–148, 2018.

[8] A. Indrzejczak. *Sequents and Trees. An Introduction to the Theory and Applications of Propositional Sequent Calculi*. Birkhäuser, Heidelberg, 2021.

[9] K. Lehrer. Induction, rational acceptance, and minimally inconsistent sets. *Minnesota Studies in Philosophy of Science*, 6:295–323, 1975.

[10] D. Leszczyńska-Jasion. *From Questions to Proofs. Between the Logic of Questions and Proof Theory*. Wydawnictwo Naukowe WNS UAM, Poznań, 2018.

[11] D. Leszczyńska-Jasion, M. Urbański, and A. Wiśniewski. Socratic trees. *Studia Logica*, 101:959–986, 2013.

[12] M. A. Lifton and K. A. Sakallah. Algorithms for computing minimal unsatisfiable subsets of constraints. *Journal of Automated Reasoning*, 40:1–33, 2008.

[13] E. D. Mares. *Relevant Logic. A Philosophical Interpretation*. Cambridge University Press, 2004.

[14] H. Omori and H. Wansing. Connexive Logics. An overview and current trends. *Logic and Logical Philosophy*, 28:371–387, 2019.

[15] R. Reiter. A logic for default reasoning. *Artificial Intelligence* 33:81–132, 1980.

[16] N. Rescher and R. Manor. On inference from inconsistent premises. *Theory and Decision* 1:179–217, 1970.

[17] G. Restall. Multiple conclusions. In D. Westerstahl P. Hajek, L. Valdes-Villanueva, editor, *Logic, Methodology and Philosophy of Science: Proceedings of the Twelfth International Congress on Logic, Methodology and Philosophy of Science*, pages 189–205. King's College Publications, London, 2005.

[18] G. Restall. Relevant and substructural logic. In D. Gabbay, J. Woods, editors, *Handbook of the History of Logic. Volume 7: Logic and the Modalities in the Twentieth Century*, pages 289–398. Elsevier/Notyh Holland, 2006.

[19] D. Scott. Completeness and axiomatizability in many-valued logic. In L. Henkin et al., editor, *Proceedings of Symposia in Pure Mathematics*, volume 25, pages 411–435. American Mathematican Society, Providence, Rhode Island, 1974.

[20] D. J. Shoesmith and T. J. Smiley. *Multiple-conclusion Logic*. Cambridge University

Press, Cambridge, 1978.

[21] T. Skura and A. Wiśniewski. A system for proper multiple-conclusion entailment. *Logic and Logical Philosophy*, 24:241–253, 2015.

[22] R. M. Smullyan. *First-Order Logic*. Springer–Verlag, New York, 1968.

[23] F. Steinberger. Why conclusions should remain single. *Journal of Philosophical Logic*, 40:333–355, 2011.

[24] Christian Strasser. *Adaptive Logics for Defeasible Reasoning*. Springer, Cham Heildelberg New York Dordrecht London, 2014.

[25] N. Tennant. Perfect validity, entailment, and paraconsistency. *Studia Logica*, 43:174–198, 1984.

[26] R. J. Wilson. *Introduction to Graph Theory. Fourth Edition*. Longman, Harlow, 1996.

[27] A. Wiśniewski. Socratic proofs. *Journal of Philosophical Logic*, 33:299–326, 2004.

[28] A. Wiśniewski. Strong entailments and minimally inconsistent sets. Research report. Technical Report 1(9), Department of Logic and Cognitive Science, Institute of Psychology, Adam Mickiewicz University, Poznań, 2016.

[29] A. Wiśniewski. Generalized entailments. *Logic and Logical Philosophy*, 26: 321–356, 2017.

www.ingramcontent.com/pod-product-compliance
Lightning Source LLC
Chambersburg PA
CBHW080451170426

43196CB00016B/2761